Cryogenics Safety Manual

Cryogenics Safety Manual
A guide to good practice

Third edition

Safety Panel
British Cryogenics Council

Butterworth-Heinemann Ltd
Linacre House, Jordan Hill, Oxford OX2 8DP

 PART OF REED INTERNATIONAL BOOKS

OXFORD LONDON BOSTON
MUNICH NEW DELHI SINGAPORE SYDNEY
TOKYO TORONTO WELLINGTON

First published by The British Cryogenics Council 1970
Second edition 1982
Third edition published by Butterworth-Heinemann 1991

© British Cryogenics Council 1991

All rights reserved. No part of this publication
may be reproduced in any material form (including
photocopying or storing in any medium by electronic
means and whether or not transiently or incidentally
to some other use of this publication) without the
written permission of the copyright holder except in
accordance with the provisions of the Copyright,
Designs and Patents Act 1988 or under the terms of a
licence issued by the Copyright Licensing Agency Ltd,
90 Tottenham Court Road, London, England W1P 9HE.
Applications for the copyright holder's written permission
to reproduce any part of this publication should be addressed
to the publishers.

British Library Cataloguing in Publication Data
Cryogenics safety manual.
   I. British Cryogenics Council Safety Panel
   536.028

ISBN 0 7506 0225 2

Library of Congress Cataloguing in Publication Data
Cryogenics safety manual: a guide to good practice/Safety Panel,
   British Cryogenics Council. – 3rd ed.
       p.   cm.
   Includes bibliographical references and index.
   ISBN 0 7506 0225 2
   1. Low temperature engineering—Safety measure. I. British
Cryogenics Council. Safety Panel.
TP482.C79 1991
621.59′028′9—dc 20

91-24934
CIP

Typeset by MS Filmsetting Limited, Frome, Somerset
Printed and bound in Great Britain

Contents

Tables and illustrations   vi
Preface   vii
Membership of Safety Panel   viii
Notice   viii
Introduction   ix

1. General safety requirements   1
   1.1 Health   1
   1.2 Safety   7
   1.3 Legislation   25
   Bibliography   26

2. Oxygen, nitrogen and argon   27
   2.1 Specific hazards   27
   2.2 Safety in the operation of air separation plants   30
   2.3 Safety in the maintenance of air separation plants   45
   Bibliography   51

3. Natural gas, ethylene and ethane   52
   3.1 Specific hazards   52
   3.2 Safety in operation   56
   3.3 Safety in maintenance   60
   3.4 Storage systems   66
   3.5 Firefighting   70
   3.6 Detection systems   74
   Bibliography   75

4. Hydrogen   77
   4.1 Specific hazards   77
   4.2 Safety in maintenance and operation of hydrogen plants   82
   4.3 Safety in maintenance and operation of liquid hydrogen storage, transport and handling equipment.   89
   Bibliography   92

5. Helium and other rare gases – research systems   94
   5.1 Helium, neon, krypton and xenon   94
   5.2 Liquid helium   98
   5.3 Grouping of neon, krypton and xenon   99
   Bibliography   100

   Index   101

Tables and illustrations

Tables

1 Thermophysical properties of various cryogens   2
2 Coefficient of expansion for various materials   12
3 Thermophysical properties of methane, ethane and ethylene   53

Figures

1 Asphyxiation hazard warning sign   6
2 Fire triangle   8
3 Toughness (impact energy) as a function of temperature for various materials   10
4 Brittle failure of pipeline   10
5 Cryogenic liquid road tanker   16
6 Cryogenic liquid rail tanker   17
7 Selection of portable cryogenic containers   18
8 Typical permit to work   20–23
9 Schematic of air separation process   31
10 LNG storage tank and bund wall   59
11 Ethylene storage tank   67
12 Liquid hydrogen storage tank   92
13 Typical liquid storage Dewar vessel and transfer tube   97
14 Phase diagram for helium   99

Preface

The main objective of the British Cryogenics Council is to promote and support the development and safe application of low temperature engineering and cryogenics in the UK.

The *Cryogenics Safety Manual* was first published by the British Cryogenics Council in 1970, and was revised in 1982. This second revision brings the manual up to date, and incorporates new material, which reflects the growth in use of cryogenic fluids and the continued importance of high standards of safety. It is intended to be a ready source of reference and provide guidance to all persons engaged in the handling of cryogenic fluids at temperatures of 188K or below, whether on a large scale in industry or research establishments, or on a small scale in laboratories.

The revision of the *Cryogenics Safety Manual* has been undertaken by a group of experts within the field of cryogenic safety which comprises the British Cryogenics Council's Safety Panel. Thanks are due to all members of the Panel for their considerable efforts, and in particular to Dr J. Currie, the Panel's Chairman. Thanks are also due to BOC Ltd, Air Products plc, British Gas plc, ICI plc, and the Institute of Cryogenics at Southampton University for making possible the formation of this team, and for providing the facilities to enable it to carry out its work.

Dr D.R. Roe
British Gas plc
Chairman, British Cryogenics Council (1988–91)

As the new Chairman of the British Cryogenics Council, I fully endorse the above remarks by Dr Roe and add my own thanks to the members of the Safety Panel. Further, I should like to mention that Dr R.N. Richardson took over as Chairman of the Safety Panel when Dr Currie moved to the USA and has been largely responsible for arranging publication of the *Cryogenics Safety Manual*.

Since cryogenics is constantly changing, the British Cryogenics Council would welcome any suggestions for inclusion in future revisions of the manual.

Professor R.G. Scurlock
University of Southampton
Chairman, British Cryogenics Council (1991–)

Membership of Safety Panel

Dr J.L. Currie (Chairman) BOC Limited
Mr P. Cook Air Products plc
Mr W. Freeman British Gas plc
Dr R.N. Richardson (Chairman 1991–) Institute of Cryogenics,
 University of Southampton
Mr R.W. Sunley ICI plc

Notice

This manual has been produced for the guidance of all who are concerned with the operation, maintenance and supervision of plant for producing, storing, and handling industrial gases at relatively low temperatures. While every attempt has been made to deal with all foreseeable hazards, the final responsibility for safety must remain with those designing, maintaining, and operating the equipment. The British Cryogenics Council accepts no liability for any errors in, or omissions from, the manual.

Introduction

The underlying principle behind all operations is to perform them safely. This is not achieved by chance but depends on detailed assessment of all the hazards. The *Cryogenics Safety Manual* provides the operator, designer and handler of cryogenic fluids and equipment with the basic safety principles that must be considered before performing any operation with cryogenic fluids.

The manual is primarily aimed at operators and managers engaged in the production, handling and use of cryogenic fluids but is recommended reading for all personnel who may become involved with cryogenic fluids. The manual attempts to cover the safety aspects of all areas of cryogenic fluid applications from large-scale industrial production to small-scale research usage.

For the purposes of the manual a cryogenic fluid is defined as one which is normally manufactured, stored, handled or processed at a temperature at or below $-85°C$ (188K).

The manual consists of five chapters. Chapter 1 describes the general safety requirements applicable to all cryogenic fluids covered in the manual. The remaining four parts look specifically at the particular hazards of defined categories of cryogenic fluids. Chapter 2 covers the atmospheric gases – nitrogen, oxygen, and argon – and deals in detail with the hazards of production and large-scale storage. Chapter 3 covers the flammable gases – liquefied natural gas (LNG), ethylene, and ethane – and again describes the safety issues in production and large-scale storage. Chapter 4 is concerned with the particular problems of handling liquid hydrogen, and how to prevent fires/explosions with this unique cryogen. Chapter 5 looks at small-scale uses of the rare gases helium, neon, krypton and xenon.

Chapter 1, 'General safety requirements', is intended to be read in conjunction with any of the remaining four parts of the manual.

This edition of the *Cryogenics Safety Manual* includes new regulations which have come into force since the previous edition and cover large-scale storage/handling of certain cryogenic fluids. The bibliography sections have also been significantly revised to include references to the latest literature.

Chapter 1 has been revised to improve the definition of those safety issues that are common to all the specific hazards of particular cryogenic fluids covered in Chapters 2–5. Chapter 3 has been restructured to reflect a similar format to the other sections. Chapter 4 has been revised to reflect modern safety thinking on the handling of liquid hydrogen. Chapter 5, on the rare gases, has been expanded to include krypton and xenon.

1 General safety requirements

A cryogen is the liquid, or under appropriate conditions solid, phase of one of the so called permanent gases; that is a gas which cannot be liquefied by the application of pressure alone at ambient temperature. In terms of thermodynamic properties this means that the critical temperature of the substance is below ambient. For the purposes of this manual a cryogenic fluid (cryogen) is defined as one which is normally manufactured, stored, handled or processed at a temperature at or below $-85°C$ (188K). The properties of various cryogens are given in Table 1.

1.1 Health

1.1.1 Cryogenic burns and frostbite

Exposure of the skin to low temperature can produce effects on the skin similar to a burn. These will vary in severity with temperature and the time of exposure.

Naked or insufficiently protected parts of the body coming into contact with very cold uninsulated pipes or vessels may stick fast by virtue of the freezing of available moisture, and the flesh may be torn on removal. Special care is needed when wearing wet gloves.

Prolonged exposure to cold can result in frostbite. There may well be insufficient warning through local pain while the freezing action is taking place. All cold burns should be checked by a first aider or, in extreme circumstances, by a medical expert to confirm extent of damage.

Prolonged inhalation of cold vapour or gas can damage the lungs. Cryogenic liquids and vapour can damage the eyes.

The low viscosity of cryogenic liquids means that they will penetrate woven or other porous clothing materials much faster than, for example, water.

1.1.1.1 Precautions

Protective clothing for handling low-temperature liquefied gases serves mainly to protect against cold burns.

Non-absorbent gloves (PVC or leather) should always be worn when handling anything that is or may have been in contact with cold liquids or

Table 1 Thermophysical properties of various cryogens

	Helium	Hydrogen	Neon	Nitrogen	Argon	Oxygen	Methane	Krypton	Xenon	Ethylene	Ethane
Chemical symbol	He	H_2	Ne	N_2	Ar	O_2	CH_4	Kr	Xe	C_2H_4	C_2H_6
Molecular weight	4	2	20	28	40	32	16	84	131	28	30
Normal boiling point °C (K)	−269 (4.2)	−253 (20.3)	−246 (27.1)	−196 (77.4)	−186 (87.3)	−183 (90.2)	−161 (111.7)	−152 (121.4)	−109 (164.1)	−104 (169.3)	−89 (184.6)
Freezing temperature °C (K)	—	−259 (14.1)	−249 (24.5)	−210 (63.3)	−189 (84.0)	−219 (54.8)	−183 (90.6)	−169 (104.2)	−140 (133.2)	−169 (104.2)	−183 (89.9)
Critical temperature °C (K)	−268 (5)	−240 (33)	−229 (44)	−147 (126)	−122 (151)	−119 (154)	−82 (191)	−63 (210)	+17 (290)	+10 (283)	+32 (305)
Critical pressure, bar	2.3	13.4	27.8	34.5	49.5	51.4	47	56	60	52.5	49.7
Expansion ratio – increase in volume as liquid at 1 bar boils to gas at 1 bar, 15°C	738	826	1417	678	820	843	626	677	556	489	437
Density of saturated liquid at 1 bar (kg m^{-3})	125	70	1200	804	1390	1142	424	2400	3100	565	546
Relative gas density (referenced to dry air at 1 bar, 15°C, density 1.21 kg m^{-3})	0.14	0.07	0.70	0.98	1.40	1.12	0.56	2.93	4.61	0.97	1.05
Latent heat of vaporisation at 1 bar (cooling potential of phase change) (h$_{fg}$ kJ kg^{-1})	23.9	451.9	87.0	199.2	162.7	212.9	512.4	108.0	96.2	483.4	488.3
Fire/explosion hazard	no	flammable	no	no	no	yes	flammable	no	no	flammable	flammable
Air liquefaction hazard	yes	yes	yes	yes	no	no	no	no	no	no	no

Note: With the exception of liquid oxygen, which is light blue, all the liquid cryogens are colourless.
With the exception of C_2H_4 and C_2H_6, which have a slight anaesthetic effect, all the cryogens listed are considered non-toxic.

vapours. Gloves should be a loose fit so that they may be readily removed should liquid splash on to them or into them.

If severe spraying or splashing is likely to occur eyes should be protected with a face shield or goggles.

Trousers should be worn outside boots and have no pockets or turnups.

1.1.1.2 First aid (cryogenic burns)

Flush the affected areas of skin with copious quantities of tepid water, but do not apply any form of direct heat, e.g. hot water, room heaters, etc. Move casualty to a warm place (about 22°C; 295K). If medical attention is not immediately available, arrange for the casualty to be transported to hospital without delay.

While waiting for transport:

(a) Loosen any restrictive clothing.
(b) Continue to flush the affected areas of skin with copious quantities of tepid water.
(c) Protect frozen parts with bulky, dry, sterile dressings. Do not apply too tightly so as to cause restriction of blood circulation.
(d) Keep the patient warm and at rest.
(e) Ensure ambulance crew or hospital is advised of details of accident and first aid treatment already administered.
(f) Smoking and alcoholic beverages reduce the blood supply to the affected part and should be avoided.

1.1.2 Oxygen deficiency (anoxia)

None of the gases listed in Table 1 have any odour, and therefore they cannot be detected by smell.

Apart from oxygen, all the gases are asphyxiants. Carbon monoxide has particularly hazardous toxic properties, which are discussed more fully in Chapter 4.

Accounts of the symptoms arising from the sudden and gradual onset of oxygen deficiency in the atmosphere are given below.

1.1.2.1 Sudden asphyxia

In sudden and acute asphyxia, such as that from inhalation of a gas containing practically no oxygen, unconsciousness is immediate. The victim falls as if struck down by a blow on the head and may die in a few minutes, unless immediate remedial action is taken.

1.1.2.2 Gradual asphyxia

Sudden asphyxia is the most common form encountered in practice but degrees of asphyxia will occur when the atmosphere contains less than 20.9 per cent of oxygen by volume.

Alexander and Himwick (1939) recognize four stages. It should be appreciated that the concentrations given are rough guides only, and may vary with individuals and ambient conditions.

- *1st stage* – oxygen reduced to 14 per cent by volume. The first perceptible signs of anoxaemia develop. The volume of breathing increases and the pulse rate is accelerated. The ability to maintain attention and think clearly is diminished, a fact that may not be noticed by the individual. Muscular coordination is somewhat disturbed.
- *2nd stage* – oxygen reduced to range 14–10 per cent by volume. Consciousness continues, but judgement becomes faulty. Severe injuries may cause no pain. Muscular efforts lead to rapid fatigue. Emotions, particularly ill temper, are easily aroused.
- *3rd stage* – oxygen reduced to range 10–6 per cent by volume. Nausea and vomiting may appear. Victim loses ability to perform any vigorous muscular movements or even to move at all. Up to or during this stage, the victim may be wholly unaware that anything is wrong. Then the legs give way, leaving the victim unable to stand, walk or even crawl. This is often the first and only warning, and it comes too late. The victim may realize that death is imminent, but does not greatly care. It is all quite painless. Even if resuscitation is possible, permanent damage to the brain may result.
- *4th stage* – oxygen reduced below 6 per cent. Respiration consists of gasps, separated by periods of increasing duration. Convulsive movements may occur. Breathing then stops but the heart may continue to beat a few minutes longer.

1.1.2.3 *Exhaustion of oxygen due to breathing in enclosed space*

It is worth noting that the effects resulting from oxygen deficiency due to addition of inert gas are very different to those which result from consuming oxygen in an enclosed space by respiration. In the latter case, oxygen absorbed is replaced by exhaled carbon dioxide. This causes discomfort and panting, thus giving early warning of oxygen depletion.

1.1.2.4 *Safe system of work – entry into confined spaces*

Oxygen-deficient atmospheres can arise from venting and purging operations. Note that inert gases which are heavier than air, e.g. argon and cold nitrogen present a special risk because of their tendency to accumulate in low places such as pits, gullies, etc.

In the UK the Factories Act 1961, Section 30, states that 'no person may enter any vessel, drain, pit, or other place where fumes may accumulate until the place has been examined and a certificate permitting entry signed by a competent person'. A shift manager is typically a competent person.

The certificate signed by the competent person may state either that the place is properly isolated, does not contain dangerous sludge and is adequately ventilated; or that it does not comply with one or more of these three requirements and breathing apparatus must be used. The wearing of a harness may also be required.

The process supervisor should prepare a permit to work, stating:

(a) The job to be carried out.
(b) The nature of any hazard present.
(c) The precautions to be taken.

(d) The results, date and time of the most recent atmosphere test.

An entry permit certificate should then be requested from a competent person.

The competent person will personally examine the equipment, and if it can be certified that the place is properly isolated and free from danger, the certificate and permit to work can be issued. If the competent person is not able to certify that the place is properly isolated and free from danger, the instructions regarding breathing apparatus and lifelines should be stated on the permit to work. The rescue plan should be rehearsed before the issue of the permit.

The certificate book should be kept in the competent person's office so that it may be inspected on request by any supervisor. The maintenance supervisor must make sure that any person working on a job covered by a certificate understands:

(a) The conditions stated on the certificate.
(b) The fact that the certificate is valid only while these conditions apply.
(c) The fact that the certificate is withdrawn automatically if there is any change in these conditions, or in the job itself or when time expires.

1.1.2.5 Rescue and first aid

Rescue personnel should ensure that they are adequately equipped with breathing apparatus, airline, etc., before attempting to remove anybody overcome in an oxygen-deficient atmosphere. Victims should be removed immediately to a normal atmosphere. If they are not breathing, it is vitally important to start artificial respiration at the first opportunity, preferably by the use of an automatic resuscitator employing oxygen gas, or alternatively by the mouth or other unaided method.

1.1.2.6 Hazard warning signs

There is no approved hazard warning sign for asphyxiant gases but the symbol in Figure 1 is generally accepted throughout the EEC and other European countries.

1.1.3 Toxicity

With the exception of carbon monoxide, the gases dealt with in this manual may be regarded as non-toxic. Abnormally prolonged exposure to some of them, however, may result in adverse effects. High concentrations, especially of hydrocarbons, may cause some nausea, drowsiness, or dizziness. Removal from exposure usually causes the symptoms to disappear rapidly.

The recommended first aid procedure for carbon-monoxide poisoning is given in Chapter 4.

1.1.4 Thermal burns

Thermal burns can arise from leaking flammable liquids and gases which have ignited, such as liquid or gaseous hydrogen, natural gas, ethane, or

Figure 1 Asphyxiation hazard warning sign

ethylene. Whereas flames from hydrocarbon gases are mostly luminous, hydrogen flames emit only very little visible radiation and are therefore difficult to see in a well-lit situation.

Maintenance of any cryogenic plant handling flammable gases should include prompt attention to product leakage.

1.1.4.1 First aid (thermal burns)

The affected areas should be treated with cold water for at least 20 minutes, and a sterile gauze dressing applied. In serious cases medical help should be summoned immediately.

1.1.5 General cold exposure (hypothermia)

Hypothermia occurs due to the body being unable to maintain its normal temperature. The dangers of hypothermia may be present at temperatures up to 10°C.

Individuals not suitably protected against low ambient temperatures may be adversely affected so far as their reactions and capabilities are concerned.

1.1.5.1 Precautions

Personnel should be protected against low ambient temperatures which may occur, e.g. on entering cold tanks or vessels.

1.1.5.2 First aid (hypothermia)

Persons apparently suffering from the effects of hypothermia should be removed from the cold area to a warm environment.

1.2 Safety

1.2.1 Safety devices

Safety devices are installed on plant and equipment to provide protection in the event of emergencies or abnormal operating conditions which may otherwise result in fire or damage. Provided the plant and equipment are operated properly, such devices will seldom be called upon to operate.

1.2.1.1 Pressure relief devices

(a) Pressure relief devices may be either spring-loaded safety valves, pilot-operated safety valves or rupture discs.
(b) The expansion ratios given in Table 1 show that warming to ambient temperature of vessels initially containing a cryogenic liquid may result in pressures between 400 and 1400 bar. Therefore vessels and pipework containing cryogenic liquids must be protected by suitable relief devices.
(c) Vents from pressure relief devices should discharge fluids to a safe area away from personnel and vulnerable equipment.
(d) All pressure relief devices should be regularly inspected for leakage and frosting. Relief valves which are leaking or 'feathering' below their set pressure will result in excessive ice formation on the valve body, which may prevent the proper operation of the valve. In extreme situations plugging of the discharge line may occur, rendering the pressure relief system inoperable.

1.2.1.2 Alarm and shutdown systems

Devices such as pressure, flow, temperature or level switches will be designed to activate audible and/or visual alarms or shut down the plant in an emergency.

1.2.1.3 Proof testing of safety devices

(a) It is suggested that unless previous experience with similar plants and environments dictates otherwise, pressure relief devices should be checked at 12-monthly intervals. The frequency of testing can then be increased or reduced depending upon operating experience.
(b) The proof testing frequency for alarm and shutdown systems should be based on the established hazard rates and reliability criteria for the plant and equipment under consideration.

1.2.2 Fire hazards

For a fire or explosion to occur, three elements are required: the igniter; the fuel; and the oxidant. These are best represented in the fire triangle (Figure 2).

The following are examples of factors leading to fire and explosion hazards.

(a) *The igniter*. Any source of heat must be regarded as an igniter in the presence of air and fuel gas. Examples are:

    (i) Naked flame.
    (ii) Lighted cigarette or pipe.
    (iii) Sparks.
    (iv) Molten metal from cutting and/or welding.
    (v) Friction.
    (vi) Static electricity.
    (vii) Electrical equipment (outside its certified area).

(b) *The fuel*
    (i) Combustible gases.
    (ii) Combustible liquids.
    (iii) Motor spirit.
    (iv) Oils.
    (v) Rags.
    (vi) Timber.
    (vii) Paper.
    (viii) Organic-based insulants.

(c) *The oxidant*
    (i) Air.
    (ii) Oxygen.
    (iii) Nitrous oxide.
    (iv) Certain acids, and their salts, eg. nitric acid, ammonium nitrate.

1.2.2.1 Emergencies

Where a fire occurs in an installation containing a flammable cryogenic fluid, the best and safest means of dealing with it is to cut off the supply of fuel to the fire. If this is not possible, it may be preferable to let the fire burn itself out. If the fire is extinguished without stopping the escape, large volumes of a

Figure 2 Fire triangle

flammable gas/air mixture may be formed, and then be ignited later by some chance source.

Information on fire precautions for specific cryogenic fluids is given in the appropriate parts of this manual. All installations with a specific fire risk should have an agreed fire fighting plan available. Personnel should be trained in its implementation at regular intervals.

Regular meetings should be held with the local fire brigade to ensure that it is aware of any special fire risks and the recommended means of countering them.

1.2.3 Oxygen enrichment

In situations where oxygen enrichment of the atmosphere can occur a potential fire hazard exists.

The atmosphere normally contains 21 per cent by volume of oxygen. Enrichment to only 25 per cent may give rise to a significant increase in the rate of combustion of burning materials exposed to such an atmosphere. Many materials, including some common metals which are not flammable in air, may burn in oxygen enriched atmospheres when ignited.

Hazards from oxygen enrichment are further explained in Chapter 2.

1.2.4 Mechanical properties of materials at low temperatures

The physical properties of many engineering materials can change significantly as temperatures fall. In particular, the toughness of materials may be considerably reduced at low temperatures and allowance must be made for this in the design of cryogenic plant (see Figure 3).

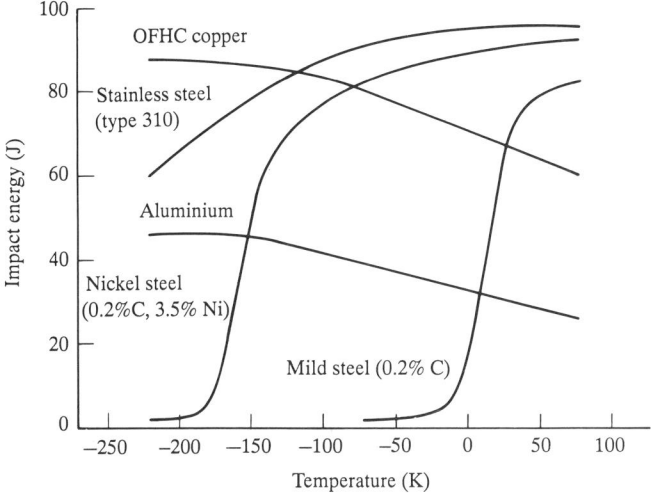

Figure 3 Toughness (impact energy) as a function of temperature for various materials. (Note: These data for Charpy V-notch impact tests have been compiled from a number of different sources and should be used for comparative purposes only)

The change in mechanical properties with reducing temperature is often very abrupt. Many materials undergo a ductile to brittle transformation at a characteristic transition temperature. Carbon and certain alloy steels are examples of metals exhibiting such behaviour. Below the transition temperature, toughness is considerably reduced, and relatively low levels of stress or shock loading may result in brittle failure. Similar behaviour may be observed in other materials, including many elastomers, polymers and composites. Even if a material does not undergo a ductile to brittle transition, repeated thermal cycling may result in stress cracking and ultimately fatigue failure. An example of brittle failure is shown in Figure 4. The carbon steel pipeline was designed for gas service but, owing to a leaking valve, cryogenic liquid leaked into the pipeline, and the resulting cooling of the carbon steel in association with an increase in pressure resulted in a catastrophic failure of the line.

When servicing or modifying cryogenic plant, it is essential that only appropriately specified components are used. In the absence of any new recommendations, or revisions of existing codes of practice which might be relevant, the original material and design specifications must be maintained. Consideration must be given to the lowest possible temperature that equipment might experience under fault conditions, and an appropriate margin of safety included in any design. It must be emphasised that cryogenic plant may not be as tolerant of abuse as systems operating at ambient temperatures.

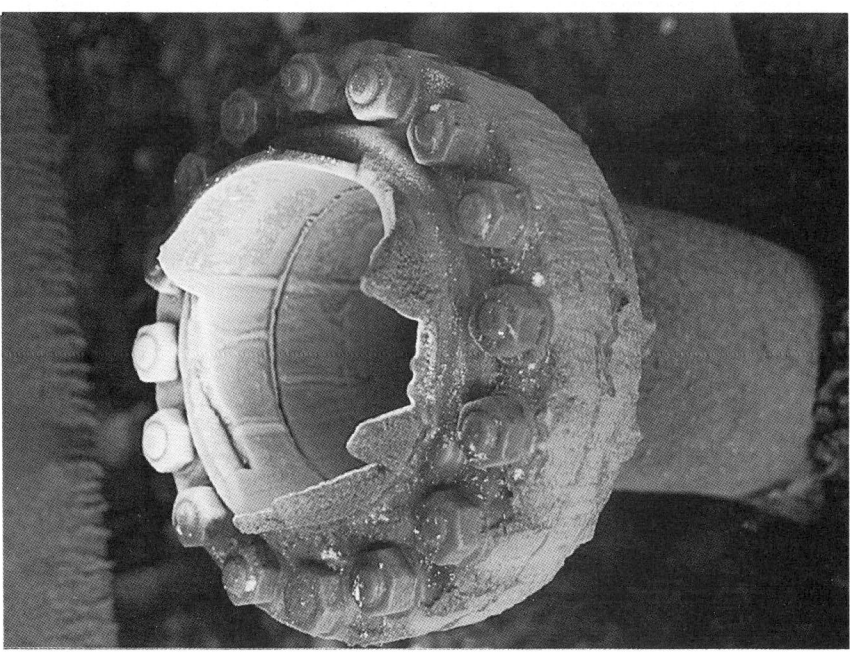

Figure 4 Brittle failure of pipeline

1.2.5 Cryogen leakage and consequent hazards

The leakage of any cryogenic substance must be considered potentially hazardous. In addition to the obvious dangers associated with the very low temperatures, the leakage of cryogens such as hydrogen and oxygen present major fire and explosion hazards. The financial implications of a leak, both in terms of cryogen loss and repair costs, must also be considered. Two main causes of leakage may be identified: leakage resulting from contraction and leakage following damage due to freezing. It should be emphasized that neither should occur in correctly designed and maintained plant.

1.2.5.1 Contraction

Most materials contract when cooled. In cooling from ambient temperature to liquid nitrogen temperature, $-196°C$ (77K), the metals commonly employed in the construction of cryogenic plant might typically contract by up to 0.3 per cent (3mm per metre). Unless such contraction has been allowed for in the design, leakage due to movement of joints may result.

Stresses due to thermal contraction can be significant and in extreme cases complete failure can occur. Fatigue failure due to thermal cycling is also possible. Coefficients of expansion (contraction) for some of the materials used in cryogenic plant are given in Table 2.

1.2.5.2 Ice formation

Water may be present in cryogenic plant as a result of condensation or liquid ingress from an external source. If an accumulation of water freezes in the pipework or vessels of the plant, the expansion that occurs during the phase change to ice may be sufficient to rupture that part of the system. This damage may remain unnoticed until cryogen is subsequently introduced to the system, with consequent loss and potential hazard. Such damage can be avoided by ensuring the plant is thoroughly dried before cool-down.

Damage may also be caused to external insulation following the ingress of either damp air or water. In addition to increasing the rate of heat inleak, such deterioration may affect the structural integrity of the insulation and result in corrosion of underlying metal, which would be difficult to detect.

1.2.6 Thermal expansion

Equipment and pipework containing a cryogenic liquid could be subject to severe overpressure due to thermal expansion of the liquid or vaporization of the liquid in a closed system. This will result from heat inleak or, in the case of liquid expansion, loss of secondary cooling.

1.2.7 Gas detection

1.2.7.1 Oxygen

Instruments used for oxygen detection employ either the paramagnetic, zirconia or electrochemical principle. The instruments may be either fixed

Table 2 Coefficients of expansion for various materials
Note: the numbers in the table must be multiplied by 10^{-5} to give coefficient α, $\Delta l = l_0 \int \alpha \, dT$.

Temperature (°C)	Aluminium	Brass	Invar (63.8%Fe, 36.0%Ni, 0.2%C)	Stainless steel (austenitic)	Pyrex glass	PTFE
−250	0	1	0	0	−1.0	37
−200	18	26	1	13	−0.5	186
−150	75	88	10	62	9	395
−100	162	168	21	120	21	691
−50	262	256	30	190	34	1216
0	370	347	47	265	49	1487
+15	430	396	53	306	56	1596

installations, for continuous gas analysis, or portable, for carrying out spot checks of atmospheres in or around various items of plant equipment.

The lightest and most portable instruments, including those of the personal pocket-sized variety, are generally of the electrochemical cell type and are battery-operated. With these types of instrument care must be taken to ensure that both the cell and battery are in good order before attempting to measure atmospheric oxygen concentration. The life of the cell as quoted by the supplier should not be exceeded and the battery should be replaced when it is approaching exhaustion. Some instruments incorporate battery test circuits.

The paramagnetic types of analyser are also available as portable instruments. Although they are usually heavier than the electrochemical cell type, they require less maintenance and can be operated from either standard mains power supply or batteries.

The zirconia types of analyser are available only as fixed instruments for continuous process monitoring.

Whichever type of instrument is used, it is important to ensure that it is well maintained and calibrated at frequent intervals. When checking atmospheres with portable analysers, it is advisable to calibrate the instrument before use; with most portable analysers calibration can be effectively carried out using atmospheric air, without the need for special calibration gas mixtures.

1.2.7.2 Flammable gas detectors

Most flammable gas detectors rely on combustion of the gas in question on the surface of a heated catalytic element. This raises the temperature of the element, and the difference between the normal running temperature of the element and that in the presence of the burning fuel gas sample is displayed on a meter, which is usually calibrated in percentage of the lower explosive limit for a given kind of gas.

Types of instrument vary. Some operate continuously in a fixed installation to give warning of excessive gas concentration at any of several measuring points; others are battery-driven and portable and only switched on when a test is required.

Alarms, either audible or visual or both, may be provided. The following precautions should be observed where flammable gas detectors are used:

(a) Ensure that the instrument in use is calibrated for the particular flammable gas which is present.
(b) Ensure that the calibration of the instrument, at the temperature at which it is to be used, is correct.
(c) To guard against false results arising from poisoning of the catalyst, the instrument should be calibrated against standard gas mixtures at regular intervals. The frequency of recalibration will be dictated by the environment in which it is used, and can only be learned by experience.
(d) If the gas sample is above the upper explosive limit, or if it contains no air (as in a tank which has been purged with nitrogen before purging with air), arrangements must be made to dilute the sample with a known percentage of air in order to obtain a valid reading. Some instruments

have a built-in facility for achieving this. If such a gas sample is not diluted with air, the reading on the detector will be zero and will lead the operator to believe, wrongly, that the atmosphere is free of flammable gas.
(e) Rapid air movement may cause gross error, and arrangements should be made for readings to be taken under calm conditions.
(f) Consult the maker regarding suitability of flammable gas detectors for use with high flashpoint materials.
(g) Check connections on sampling lines to ensure that there is no undesirable dilution of the sample.
(h) Ensure that flammable gas detectors comply with appropriate current British Standards, e.g. BS6020 and BS5501.
   Certification of the instrument for the gases with which it is used should be in accordance with BASEEFA Standard SFA 3007:1981 'Instruments for measuring gas concentration', and should be obtained from the manufacturer.
(i) In oxygen-rich atmospheres, normal certification is invalid, and the calibration made with air may not be correct.
(j) An atmosphere containing a large percentage of water vapour may change from 'safe' to 'explosive' if the water vapour condenses.
(k) Ensure that the operation and maintenance instructions and precautions given in the maker's handbook are strictly followed. Ill-maintained, carelessly used equipment tends to give a low, rather than a high, reading.

I.2.7.3 Toxic gas detectors

These, by definition, cover a broader range of compounds than those described in the two preceding sections. The most common toxic gas encountered in cryogenic gas separation plants, e.g. hydrogen separation plants, is carbon monoxide. Monitoring and detection of toxic gases in the atmosphere can be achieved on a continuous basis by permanently installed process analysers linked to gas alarm systems, or on an intermittent basis by using portable detectors. The latter type should be used with caution, depending upon the risk and degree of exposure of the operator to toxic gases.

Portable detectors can be either battery-operated instruments or in the form of reagent tubes. The reagent tube is the cheapest method of detection and is generally reliable as a means of obtaining a quick, approximate indication of toxic gas concentration.

The reagent tube consists of a graduated glass tube filled with a solid reagent which changes colour on exposure to the toxic gas. The atmosphere to be tested is aspirated through the tube by a manually operated pump at a controlled rate. The number of times the pump is operated defines the volume of sample passing through the reagent. The length of the reagent which undergoes a colour change gives an estimate of the concentration of toxic gas in the atmosphere.

The following precautions should be used when operating this type of detector:

(a) Ensure that the tube is a gastight fit in the pump inlet connection.

(b) Follow carefully the maker's instructions regarding the operation of the pump.
(c) Follow carefully the maker's instructions regarding shelf life and storage conditions for the tubes.
(d) Use a new tube for each measurement.
(e) Dispose of used tubes safely. Many contain extremely reactive chemicals.

1.2.8 Distribution of cryogens

Cryogenic liquids are stored and distributed using purpose designed equipment for each product. Transport of large quantities of cryogen may be by road tanker, rail tanker, ship or pipeline. Smaller quantities are conveniently handled in portable Dewars, liquid cylinders and other insulated vessels. The design and construction of the equipment used to transport cryogens takes into account the physical and chemical properties of the specific products. In all cases distribution must be accomplished in complete safety and with minimum loss of cryogen.

1.2.8.1 Road and rail tankers

These are relatively sophisticated vehicles. Drivers and operators must receive formal training in correct operating procedures.

Procedures must be established to prevent the possibility of product contamination or loading of the wrong product. This is particularly important where flammable or oxidizing products are concerned.

All road and rail tankers must be clearly identified with the product name and carry appropriate safety and emergency instructions. A typical cryogenic liquid road tanker is shown in Figure 5 and a rail tanker in Figure 6.

1.2.8.2 Pipelines

Insulated pipelines may be used to convey cryogens over distances from a few metres to several hundred metres. The maximum distance will depend on the product and the quality of insulation. Although a solid insulant is most commonly employed, increasing use is being made of vacuum insulated pipelines. The general safety considerations relating to cold surfaces and insulation breakdown are obviously applicable, but the operation and maintenance of a pipeline is a specialized undertaking and in every case the designated operating procedures must be strictly followed.

1.2.8.3 Dewars and insulated vessels

Insulated containers for the storage and distribution of smaller quantities of cryogens range from simple open top vessels moulded in polystyrene, which are designed to hold a few litres of cryogen (primarily nitrogen) for short periods of time, to complex vacuum insulated Dewars with capacities of several hundred litres, which offer extremely low rates of boil-off. See Figure 7.

Moulded vessels are inexpensive but suffer relatively high rates of boil-off. These containers are quite fragile and should be inspected for cracks before use. The only significant hazard associated with this type of container is the ease with which they may be knocked over, particularly when only part full, owing to their light weight.

Vacuum insulated Dewar vessels are manufactured in a range of sizes up to about 200 litres. The vessels may be constructed from metal, glass, composites or any combination of these. Many designs include additional materials in the vacuum space to enhance the quality of insulation. Dewars designed for hydrogen and helium may use a liquid nitrogen-cooled outer vessel.

The majority of Dewars are not designed to operate at pressures significantly above or below ambient. Blockage of the neck of a Dewar or a loss of vacuum (and hence insulation) can lead to an increase in internal pressure, resulting in catastrophic failure. Partial loss of vacuum may also result in the ingress of atmospheric air, which may condense in the vacuum space. When the Dewar subsequently warms up, the trapped liquid air will evaporate and may not be able to escape fast enough to prevent an increase in pressure sufficient to burst the vacuum shell.

In the case of helium and hydrogen Dewars care must be taken to prevent air diffusing down the neck, where it can freeze to form a plug. In Dewars containing the higher boiling point cryogens, such as nitrogen, oxygen or LNG, ice plugs forming from atmospheric water vapour are the major danger. All Dewars should have some form of pressure relief system, but this

Figure 5 Typical cryogenic road tanker (courtesy Air Products plc)

General safety requirements   17

should not be relied upon. In the event of a Dewar becoming blocked the manufacturer's instructions should be followed.

Fortunately the rate of pressure increase within a blocked Dewar is usually quite slow, provided the insulation is maintained. One technique frequently employed to relieve the pressure is to slowly melt a small hole through the plug, using a hot copper rod or tube. The tube is to be preferred, since it then provides a vent path when the plug is finally breached.

1.2.8.4 Liquid cylinders

Liquid cylinders are a special type of Dewar designed to deliver a gaseous product at moderate pressures. At present their use is restricted to oxygen, nitrogen and argon but there is no practical reason why suitably designed liquid cylinders should not be used for other cryogens. Transport and storage of a product in liquid form has many advantages over the use of high pressure gas cylinders. Liquid cylinders have inbuilt evaporator and pressure raising circuits. The cylinders are designed to cope with fault conditions, with excess pressure being vented via safety valves. It is important that the safety devices are inspected regularly and that the cylinder is only operated within its rated capacity. In common with all Dewars, liquid cylinders should always be handled with care and not subjected to any undue shock or loading.

Figure 6 Cryogenic liquid rail tanker (courtesy BOC Limited)

1.2.9 Safety control procedures

1.2.9.1 Safety work permits

There will be occasions when it will be necessary, either on a planned or emergency basis, to carry out work either on or in the vicinity of a gas-processing plant which may call for inspection, modification, maintenance, or repair, and which usually requires a departure from normal production routine. Under these conditions plant maintenance personnel or contract workers may be exposed to danger purely as a result of ignorance of the potential hazards which may exist. Such hazards, which have already been described, include pressurized systems, low temperature fluids, and flammable, asphyxiating or oxygen-rich atmospheres.

If accidents are to be prevented, it is vital that the closest possible understanding exists between those responsible for production and those carrying out the repair and maintenance work. Experience has shown that the maximum degree of safety can be achieved by using a 'permit to work' system as a formal and disciplined procedure. This requires that a written statement be prepared and signed by a responsible person for issue to the individual in charge of the maintenance operation, to the effect that the particular section of plant or piece of equipment is safe for work to begin.

The plant manager should be responsible for ensuring that procedures for operating the work permit system are established and that they are understood and adhered to by all personnel and contractors engaged in the maintenance operation. An example of a 'work permit' is shown in Figure 8.

Figure 7 Selection of portable cryogenic containers

1.2.9.2   Checklists

An effective method for ensuring that plant and equipment have been made safe for work to be carried out and that the appropriate safety precautions will be observed is the use of a checklist. Such a list may either be built into the work permit itself or used separately as a reminder to those persons responsible for issuing work permits.

Although checklists will not cater for every conceivable hazard, it is possible to enumerate the fundamental actions which must be taken to make equipment safe but which can easily be forgotten under pressure of work. The principal items which should be included in a checklist are the following.

1.2.9.2.1   Actions to make plant and equipment safe

(a) A review of the process flow sheet and of the actual area and equipment where the work is to be carried out.
(b) Effective isolation of equipment from other parts of the plant or process which may still be 'live' and which could create a hazard by inadvertent operation of a

    valve, switch, or control system. The checklist shall carefully enumerate the isolation measures to be taken such as:

    (i) Stop valves.
    (ii) Double block and bleed.
    (iii) Blank flanges or slip plates.
    (iv) Locks and/or signs.

(c) The effective isolation of the power supply from prime movers, e.g. electric motors, steam turbines.
(d) Effective depressurization of equipment and provision of adequate arrangements to ensure that it remains depressurized during the entire maintenance operation.
(e) Normalization of temperature, to ensure that personnel are not exposed to excessively high or low temperatures.
(f) Removal of flammable, toxic, or otherwise harmful fluids and materials from process lines, equipment and working environment.
(g) Analysis of surrounding atmosphere to check on any change of the oxygen content, or existence of explosive or toxic mixtures. This requirement is of prime importance when it is necessary for workers to enter process vessels or equipment, or if hot work is to be carried out.
(h) Specification of requirements for the carrying out of periodic analyses of the atmosphere. It will be necessary that production staff are aware of the acceptable limits for safe working.
(j) A clear statement as to whether work requiring naked flames or other sources of ignition is permitted.
(k) Provision of emergency arrangements, e.g. fire extinguishers, means of escape, standby personnel.
(l) A clear statement of any special precautions that may be necessary.
(m) A reminder that the specified and special safety precautions required do not release the personnel carrying out the work from the responsibility of taking general precautions (signs, area roped off, manholes covered, etc.).

PLANT			EQUIPMENT
	PART A	**PREPARATION**	

A.1   WHILE DOING THE JOBS DETAILED IN B.1 THE FOLLOWING HAZARDS MAY BE MET (SPECIFY PRECAUTIONS TO BE OBSERVED IN B.2)

Strike out those for which there will be no hazard specific to the job when preparation is complete	**Asbestos** **Corrosive, hot and other liquids** **Created openings** **Dust** **Electricity – mains, static** **Fire and explosions** **Gas, fumes** **Gas or liquid under pressure** **Hot metal** **Hydraulic pressure**	**Moving machinery** **Noise** **Other jobs nearby** **Overhead hazards – cranes, pipes, cables, etc.** **Radioactive substances** **Steam** **Toxic materials** **Trace heating** **Traffic (road and rail)** **Underground services**

ADD OTHER HAZARDS ....

A.2   Strike out as necessary

ISOLATION (other than electrical)
Physical isolation is NOT APPLICABLE
The place, equipment, etc, is NOT ISOLATED
Equipment etc. was isolated on P.T.W.
The place, equipment, etc. is ISOLATED

METHOD OF ISOLATION delete as appropriate:
SINGLE/DOUBLE ISOLATION VALVE CLOSED AND LOCKED OFF
LINES SLIP PLATED
PHYSICAL DISCONNECTION–OPEN END BLANKED OFF
VENT, DRAIN OR BLOW OFF OPEN

OTHER INFORMATION
....
....

A.3   **PRECAUTIONS ALREADY TAKEN**
....
....

A.4   Strike out ONE Statement

FIRE PERMIT
A Fire Permit is NOT NECESSARY
A Fire Permit IS NECESSARY   permit Issued No. ....
Further Fire Permits can be noted on the back of this PTW

A.5   Strike out ONE Statement

FACTORIES ACT 1961, Section 30, and CHEMICAL WORKS REGULATION 7
DO NOT APPLY
DO APPLY – I have seen the signature of the responsible person

on Permit to Enter No. ....
Further Entry Permits can be noted on the back of the PTW

A.6   Strike out as Necessary

INSTALLED RADIOACTIVE SOURCES
There is NO INSTALLED RADIOACTIVE SOURCE
There is AN INSTALLED RADIOACTIVE SOURCE and it was made safe on

PTW No. ....
There is AN INSTALLED RADIOACTIVE SOURCE and the Safety Instruction for the installation DOES REQUIRE/DOES NOT REQUIRE it to be isolated before work is done (see inside cover for definitions).
I have made the installation safe for the duration of the Permit by
....
Signature of person ....

A.7   **ELECTRICAL ISOLATION** (ALL isolations are to be recorded by signature)
ELECTRICAL ISOLATION IS APPLICABLE/IS NOT APPLICABLE
HIGH VOLTAGE PERMIT TO WORK CARD NO. _____

Item/circuit reference.						
Fuses out						
Locked off						
Racked out						
Start button tried						

A.8   Strike out as Necessary

MASTER CONTROL SHEET

Master Control Sheet No. .... applies

A.9   **PREPARATION COMPLETE**

OR 1626   (print   **Signature** ....

Figure 8   Typical Permit to Work

EQUIPMENT No.

PART B	OPERATION	
B.1	**JOBS TO BE DONE**	Tag Nos.

Equipment Rotation Check: NECESSARY/NOT NECESSARY
(To be carried out in accordance with Plant Instructions)

B.2	**PRECAUTIONS TO BE TAKEN AND WHY**

(Strike out those not required and enter others in the spaces below)

PROTECTIVE CLOTHING/EQUIPMENT TO BE WORN
(Specify type if not standard issue)

PVC suit
PVC gloves
Hearing protection
Helmet
Ori-nasal mask

Full face shield
Goggles
Beware trapped pressure/liquids –
wear gloves and goggles for first break in

OTHER PRECAUTIONS THAT RELATE TO THE HAZARDS IN A.1

ASBESTOS

Insulation materials used in the work area may contain some asbestos and personal care must be exercised to avoid disturbance.

Figure 8 *continued*

VALID UNTIL: DATE   **TIME:**

PART C   **ISSUE, ACCEPTANCE AND RETURN**

THIS PERMIT IS ONLY VALID WHEN ALL SECTIONS (A1 TO C1) HAVE BEEN COMPLETED

C.1 ISSUE AND ACCEPTANCE

	Name (Printed)	Signed or Initialled	Date–Time
I have read conditions on this Permit	FROM	....	....
	TO	....	....

C.2 ENDORSEMENT BY OTHER(S) ENGAGED IN SUPPORT ACTIVITY OR CO-ORDINATION ROLE RE THE PRIME ACTIVITY DETAILED IN B.1

Signing On	(1)	(2)	(3)	(4)
Trade				
Name (Printed)				
Signed or Initialled				
Tel. No.				
Date				

(Where daily endorsements required use the back of this Permit)

Signing Off
Signed or Initialled
Date

C.3 TEST RUN/ROTATION CHECK REQUIRED

Signatures (from C.1) Issuer: ....   Acceptor: ....

Electricity restored   Signed: ....

TEST RUN COMPLETED/ROTATION CHECK CORRECT (delete as necessary)

Signatures (from C.1) Issuer: ....   Acceptor: ....

Electricity isolated   Signed: ....

C.4 COMPLETION OF JOB

Strike out ONE Statement
- THE JOB IS COMPLETE
- THE JOB IS INCOMPLETE and a further Permit to Work No. .... is prepared for the following work:

....
....

RETURN   Signatures   By .... Date–Time ....
                       To .... Date–Time ....

C.5 MASTER CONTROL SHEET EXAMINATION

Strike out as Necessary
Isolations on Sheet No .... MUST REMAIN
Isolations on Sheet No. .... MAY BE REMOVED
Signature .... Date ....

C.6 FACTORIES ACT 1961, Section 30, and CHEMICAL WORKS REGULATION 7

If these apply the following declaration is to be completed by the person taking back the plant. I have notified Mr. .... the person currently responsible for Permit to Enter No. .... that this PTW is now being withdrawn

Signature .... Date ....

C.7 INSTALLED RADIOACTIVE SOURCES (Delete if not applicable)

Strike out as necessary
- The radioactive source remains ISOLATED
- The radioactive source is REMOVED
- The radioactive source has been RECOMMISSIONED

Signature of qualified person ....

C.8 ELECTRICAL SUPPLY Refer to Isolations A7

Item/circuit Reference				
Electricity Restored				

Electrical Isolations Remain, ...

Transferred to PTW No. .... Signed ....

Figure 8 *continued*

DAILY OR SHIFT ENDORSEMENT IN ACCORDANCE WITH SITE INSTRUCTIONS

I have read the conditions on this Permit (overleaf) ....

DATE	FIRE PERMIT NO.	ENTRY PERMIT NO	PROCESS	TRADE 1	TRADE 2	TRADE 3	CONTRACTOR

Figure 8 *continued*

1.2.9.2.2 *Protective clothing and equipment*

(a) Safety glasses.
(b) Chemical apron.
(c) Safety harness.
(d) Dust mask.
(e) Fresh air masks.
(f) Hard hat.
(g) Chemical goggles.
(h) Chemical gloves.
(j) Lifeline/wrist strap.
(k) Spark proof tools.
(l) Fire retardant clothing.
(m) Oxygen analyser.
(n) Faceshield.
(o) Ear protection.
(p) Warning horn.
(q) Low voltage lights.
(r) Explosimeter
(s) Self-contained breathing apparatus.
(t) Clean, oil free overalls/overshoes.

1.2.9.3 *Special precautions*

Although a general purpose permit system will be adequate for most situations which are liable to be encountered, it may be advisable to introduce additional permit systems to cover specialized areas, examples of which are:

(a) High voltage electrical equipment.
(b) Entry into vessels or confined spaces (see 1.1.2.4).
(c) Work requiring the use of flames or other potential sources of ignition.

1.2.10 Rapid changes in operating parameters

When plant handling or processing cryogenic liquids is started up or shut down, rapid changes occur within the system. Rapid changes of flow, pressure, temperature or composition are often the precursor to the realization of the hazards described in this section. It is important that all personnel involved in the design, operation and maintenance of cryogenic liquid systems recognize this fact.

1.3 Legislation

1.3.1 Major hazards

The storage of large quantities of cryogenic flammable gases, toxic gases and liquid oxygen is covered by specific safety requirements in European and UK legislation.

An EC Council Directive (82/501/EEC) on the 'Major Accident Hazards of Certain Industrial Activities' was passed in 1982, with further amendments in March 1987 and November 1988 which set the standard for major hazard legislation in EC countries. In the UK the EC Directive was implemented by the Control of Industrial Major Accident Hazards (CIMAH) Regulations 1984 and subsequent amendments.

In certain circumstances installations handling cryogenic liquids will be subject to the Notification of Installations Handling Hazardous Substances Regulations 1982 (NIHHS) and the Control of Industrial Major Accident Hazards Regulations 1984 (CIMAH).

The CIMAH Regulations require that persons in control of an industrial activity that is subject to the Regulations shall at any time be able to demonstrate that the activity is being operated safely, and they place an obligation on such persons to report major accidents (regulations 4 and 5). More stringent requirements contained in regulations 7 to 12 (known as 'top-tier' requirements), which apply to only the potentially more hazardous activities, call for the following:

(a) A safety report to be sent to the Health and Safety Executive (HSE) at least 3 months before commencement of the activity.
(b) The preparation of an on-site emergency plan before commencement of the activity.
(c) Provision of information to the local authority (at county or equivalent level), to enable them to prepare an off-site emergency plan.
(d) Distribution of safety information to the public before commencement of the activity.

The top-tier requirements of CIMAH apply to industrial installations carrying out a processing operation or storage of a cryogenic liquid above the quantity shown below. In assessing the quantities, account should be taken of the substance in on-site storage and on-site transport associated with the process operation, as well as of the quantities in the process itself.

	Quantity (tonnes)
Liquid oxygen	2000
Liquid natural gas	200
Liquid ethylene	200
Liquid ethane	200
Liquid carbon monoxide	200
Liquid hydrogen	50

The above quantities may be subject to change, and the reader is advised to check the regulations for the current specified quantity.

1.3.2 Planning controls

Planning restrictions will almost certainly apply to locations where storages of the quantities covered by the CIMAH regulations are proposed. The storage of smaller quantities of cryogenic flammable gases, toxic gases and liquid oxygen may also be subjected to planning control under the Housing and Planning Act 1986 (Part IV).

1.3.3 Dangerous Substances (Notification and Marking of Sites) Regulations 1990

Sites containing 25 tonnes or more of dangerous substances must:

(a) Erect suitable signs at access points to warn those entering the site of the existence of those dangerous substances. Compliance with this provision is required from 1 September 1990.
(b) Notify the appropriate fire and enforcing authorities, giving information specified in the regulations. Compliance with this provision is required from 1 October 1990.
(c) Erect suitable signs at individual locations on site warning of the presence and nature of the hazard of those dangerous substances, as directed by an inspector.

The 25 tonne threshold is an aggregate total.

The cryogenic liquids listed above for the CIMAH regulations are defined as dangerous substances for the purposes of these regulations.

Bibliography

Alexander, F.A.D. and Himwick, H.E., *American Journal of Physiology*, 1939, 126, p. 418
British Cryogenics Council Symposium, 'Safe storage and handling of cryogenic liquids', London 8 March 1988. Papers published in '*Cryogenics*' November 1988, Butterworth Scientific Ltd, Guildford, Surrey.
BS 5429, Code of practice for safe operation for small scale storage facilities for cryogenic liquids (1976)
European Industrial Gases Association, Brussels. Work Permit Systems, Industrial Gases Council Doc. 40/90.
Hands, B.A., *Cryogenic Engineering*, Academic Press, 1986.
Pressure Systems and Transportable Gas Containers Regulations, The (1989), Statutory Instrument No. 2169.
'Using Liquid Nitrogen in the Engineering Industries', BCC Conference Papers, 1990.
Webster, T.J., 'When Safety is Dangerous', *Proceedings of 2nd International Symposium on Loss Prevention and Safety Promotion in the Process Industries*, Heidelberg, September 1977.
Webster, T.J., 'Industrial Gases' *Royal Insurance Journal*, Hazard, Autumn 1978, No.17, pp. 22–29
Zabetakis, M.G., *Safety with Cryogenic Fluids*, Heywood Books, London

2 Oxygen, nitrogen and argon

Chapter 2 deals with special precautions which must be observed in the operation and maintenance of air separation plants and product handling and storage equipment. For the purpose of this guide an air separation plant includes all machinery and equipment for the production of oxygen, nitrogen, or argon by the purification and distillation of liquid air. Handling and storage equipment refers to all pumps, compressors, vessels, and ancillary equipment in service with the gases and liquids produced by the air separation plant. This chapter supplements the general safety requirements of Chapter 1.

2.1 Specific hazards

2.1.1 Oxygen

2.1.1.1 Fire and explosion hazard

Oxygen is not itself a flammable gas, but vigorously supports combustion. Combustible materials ignite more easily and burn more rapidly in an atmosphere which contains a higher concentration of oxygen than that of air, i.e. 20.9 per cent by volume. These effects intensify with increase of oxygen concentration, pressure and temperature. Many commonly used materials not normally combustible in air may burn in pure oxygen or oxygen-enriched atmospheres. Special precautions are therefore essential when using equipment in service with either gaseous or liquid oxygen if fires or explosions are to be avoided.

2.1.1.2 Selection of materials for oxygen service

Only materials which have been properly tested and approved may be used in service with oxygen. Information regarding the selection of such materials should be obtained either from the industrial gas suppliers or published data.

The behaviour of several common substances in contact with gaseous or liquid oxygen is summarized below. The list is not exhaustive, but serves as an indication of the types of materials which may be safely used in oxygen service and those which are not acceptable for use.

(a) *Materials that are highly dangerous*
Flammable liquids and gases.
Porous flammable materials that will absorb oxygen.
Bitumen based substances.
Hydrocarbon-based oils and greases.

(b) *Materials that are unsafe (will ignite easily)*
Wood.
Asphalt.
Paint.
Clothing materials
Cotton waste.
Finely divided metals and carbon.
Cork.
Organic solvents.

(c) *Materials that will normally be safe (will not ignite easily)*
Stainless steels.
Mild steel.
Cast iron and cast steel.
Aluminium.
Aluminium bronzes.
Zinc.
PTFE.
Note. Although these materials are normally safe for approved applications, they can ignite under some conditions, particularly in the finely divided state, and continue to burn.

(d) *Materials safe under practically all conditions (will not normally ignite in solid form)*
Copper.
Brass.
Bronze.
Gold.
Silver.
Nickel.
Monel and other non-ferrous nickel alloys.
Oil-free silicate-based insulants.

2.1.1.3 Hazards to personnel

Oxygen is odourless, colourless, and tasteless; therefore atmospheres enriched in oxygen cannot be readily detected by the normal human senses. In general, oxygen should never be released into confined spaces where there is inadequate ventilation. Liquid oxygen and gaseous oxygen are heavier than air and can accumulate in low-lying areas, such as pits and trenches, where the gas may be slow to disperse. Liquid oxygen can also migrate and seep through porous materials, fissures in the soil, cracks in concrete and roadways, drains, and ducts.

Personnel should not enter areas where oxygen enrichment of the atmosphere is suspected without first testing the atmosphere with a suitable gas detector to ensure that the oxygen content lies between 20 and 22 per cent by volume. The principal danger is that body hair, clothing and porous substances may become saturated with oxygen, whereupon they will burn violently if ignited.

Clothing which has been contaminated with an enriched oxygen atmosphere should be well ventilated in the open air for about 15 minutes before the wearer approaches a source of ignition.

2.1.1.4 Misuse of oxygen

It is extremely dangerous to use oxygen gas as a substitute for compressed air, nitrogen or other gases. Oxygen should never be used for any of the following or similar applications:

(a) Starting internal combustion engines.
(b) Pressurizing oil reservoirs.
(c) Operating pneumatic tools.
(d) Paint-spraying.
(e) Inflation of vehicle tyres.
(f) Blowing out pipelines, etc. (other than oxygen service lines, which may be purged with oxygen).
(g) Freshening the air or clearing the fumes in a confined space.

The Health and Safety Executive leaflet 'Fires and Explosions due to Misuse of Oxygen', No. 8 1984, contains further information on this hazard.

2.1.2 Argon and nitrogen

2.1.2.1 Asphyxiation hazards

Both argon and nitrogen are colourless, odourless, and tasteless, and cannot be detected by the normal human senses. They are non-flammable, but act as asphyxiants by displacing the oxygen from the atmosphere. Argon and cold gaseous nitrogen are heavier than air and may accumulate at low points in pits and trenches. Before entering areas or equipment suspected of being deficient in oxygen, test the atmosphere with an oxygen detector. Personnel should not be exposed to atmospheres containing less than 20 per cent by volume of oxygen. The BCC has prepared a safety package, 'Hazards with Nitrogen and Argon', which covers this subject in more detail.

2.1.2.2 Fire and explosion hazards

Oil-lubricated compressors operating continually on nitrogen or argon service for a prolonged period should not be switched to air service without thorough cleaning; otherwise there is a danger that unoxidized pyrophoric deposits which may have formed in the machine will explode violently on contact with compressed air.

At equal pressures, the boiling point of liquid nitrogen is lower than that of liquid air. Air will condense on the external surfaces of vessels or pipework containing liquid nitrogen at an equilibrium pressure less than 1.5 bar absolute if the vessels are either unlagged or lagged with a porous cellular type insulant which has not been properly vapour-sealed. The liquid air produced can result in oxygen enrichment of the atmosphere local to the equipment, and if the insulant is combustible, there is a serious risk of fire. Special care must therefore be taken before any maintenance or repair work is started, particularly where the use of open flames or other potential sources of ignition is intended.

2.2 Safety in the operation of air separation plants

The operating instructions provided by the plant manufacturer for an air separation plant will usually contain the general safety requirements, together with the particular safety precautions relating to specific operating procedures. It is vital that the plant management and operating personnel both understand and familiarize themselves with all safety requirements and adhere to them in practice.

The basic operation of an air separation plant includes the liquefaction of air inside an insulated enclosure, a cold box. By rectification and distillation it is separated into its main constituents nitrogen, oxygen and argon. The liquid components are either stored at cryogenic temperatures in thermally insulated low-pressure tanks or evaporated and compressed into pipelines for onward distribution.

The purpose of this section of the manual is to emphasize those areas of operational safety which are of fundamental importance and generally applicable to all air separation plants and handling and storage equipment.

2.2.1 General

2.2.1.1 *Housekeeping*

For reasons of general safety and because of the hazards associated with oxygen, it is important to ensure that high standards of housekeeping and cleanliness are maintained throughout the entire factory.

Storage of equipment, spare parts, and potentially hazardous materials such as oil and grease should be located as far away as practically possible from the plant operating area, particularly the sections of plant handling the gaseous and liquid products. A separation distance of not less than 15 metres is recommended for potentially hazardous materials.

2.2.1.2 *Prohibition of smoking and open flames*

Smoking and the use of open flames are forbidden in the vicinity of the plant while it is in operation or contains liquid or gaseous oxygen. It is the responsibility of the management to define:

(a) The areas where smoking is prohibited.
(b) The areas and conditions under which open flames, e.g. cutting or welding operations, may be used.

Where maintenance or repair work using heat or open flames is to be carried out in the plant operating area, the use of a 'permit to work' system is essential.

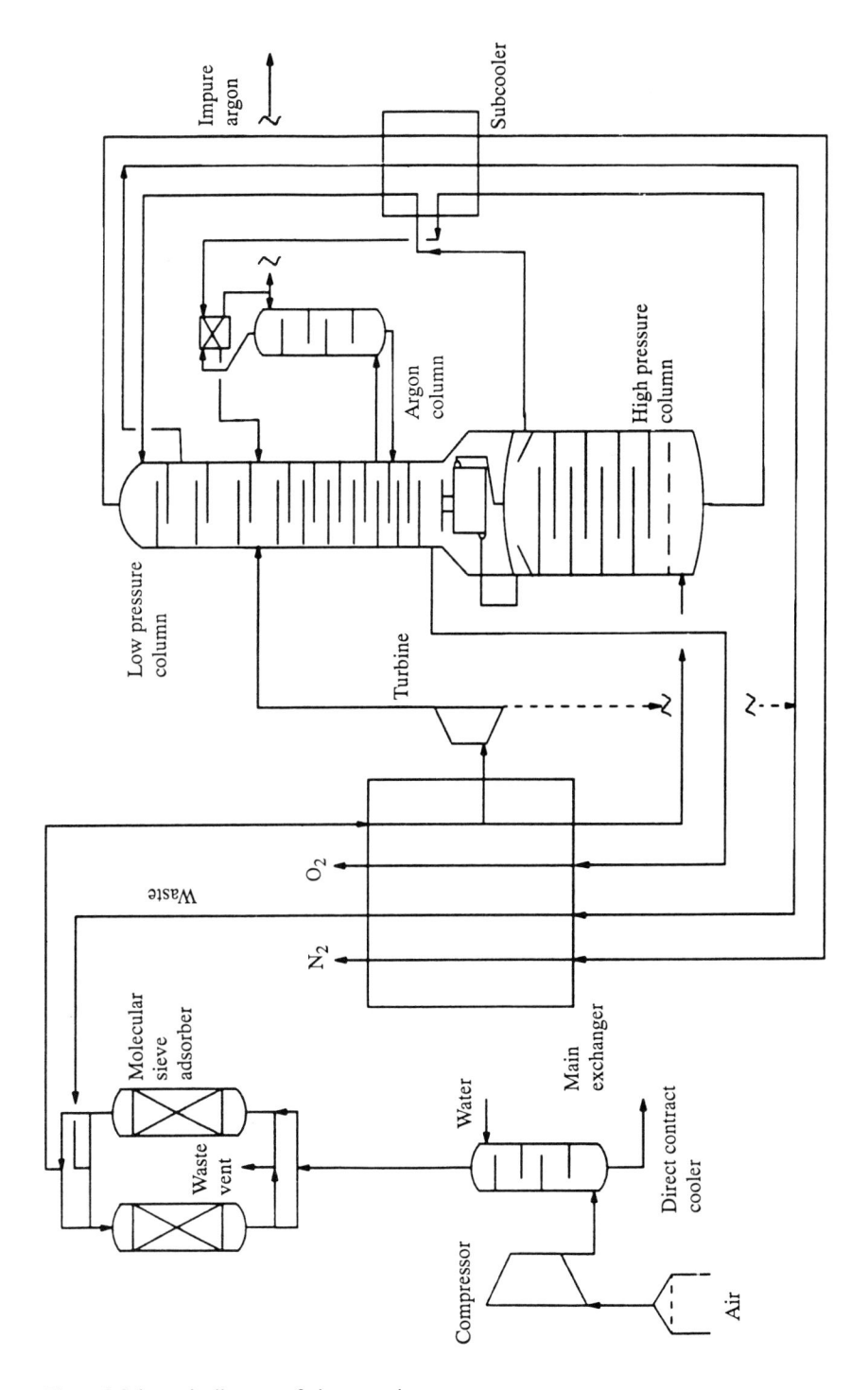

Figure 9 Schematic diagram of air separation process

2.2.2 Contaminants in the process stream

Oxygen, because it can react with many substances, must be manufactured and stored in equipment which is kept free of contamination. Where contaminants cannot be completely eliminated, sound design and safe operating practices enable them to be controlled within acceptable limits.

Contaminants may be introduced into air separation plants by way of the air-feed stream, by plant equipment malfunction, or inadvertently during maintenance or construction.

2.2.2.1 Control of airborne contaminants

A knowledge of the nature and concentration of the contaminants in the process air is of fundamental importance in the safe operation of the plant.

Non-flammable contaminants, such as carbon dioxide and water, are always present, but do not normally create safety problems unless, through plant maloperation or process instability, carbon dioxide passes through to the low pressure column and plugs the reboiler passages. Under these circumstances local dry boiling and concentration of potentially hazardous contaminants, such as acetylene, can occur.

Particulate matter is normally removed from the air-feed stream by filters located on the suction side of the main air compressors.

The presence of corrosive gases, such as sulphur dioxide or oxides of nitrogen, in the atmosphere can, over a period of time, cause severe corrosion of the warm end sections of the plant unless proper controls are introduced. Periodic monitoring of the condensate from the aftercooler of the air-feed compressor for pH level will serve as a useful indication of the degree of acid-gas contamination in the atmosphere.

In severe cases it may be necessary to provide cleaning or scrubbing equipment to reduce the acid-gas concentration in the air before it enters the feed compressor.

The contaminants of greatest concern are those which are flammable, such as hydrocarbons, and which can pass through to the oxygen-rich sections of the process system. Acetylene is of particular concern, because of its relatively low solubility and high reactivity in liquid oxygen.

Although air separation plants normally incorporate equipment for removal of hydrocarbon contaminants from the gas and liquid streams, the design of which is based on measured or estimated pollution levels, it is advisable to be aware of local sources of contamination which may on occasions give rise to peak levels and jeopardize the safety of the plant. Common sources of environmental contamination are:

(a) Chemical and oil refining processes.
(b) Exhausts from internal combustion engines.
(c) Acetylene generating equipment.
(d) Other local industries or processes.

Air separation plants are frequently located in industrial areas where a degree of atmospheric contamination will always be present. The initial selection of the plant location is therefore of great importance in minimizing the levels of undesirable materials in the air-feed stream.

It is of the utmost importance that correct operating procedures are maintained to ensure that the levels of contaminants in the various sections of the plant are controlled within acceptable limits. Particular attention should be paid to the following items.

2.2.2.1.1 Control of plant regenerator/heat exchanger temperatures. For plants fitted with regenerators or reversing heat exchangers it is important to control the cold-end temperatures below the maximum safe temperatures stated in the working instructions. The regenerators or reversing heat exchangers provide an important safeguard in removing the heavier hydrocarbons from the incoming air, and this function will not be performed effectively if the cold-end temperatures rise too high.

If the cold-end temperature rises excessively, due to maloperation, faults in the switch valves or switch valve timing sequence, or other plant upsets, or if it is too high in any regenerator when the plant is restarted in the cold condition, the heavier hydrocarbons which are normally purged to atmosphere by the waste nitrogen cycle may be carried into the separation column in high concentrations during the air cycle.

If at any time the cold-end temperatures of a regenerator or reversing exchanger rise above the limits given in the working instructions, immediate action should be taken to restore them to their correct operating levels.

2.2.2.1.2 Purging of air separation plant. Where the plant working instructions call for certain drain valves to be blown at specified intervals, or for continuous liquid purging, these instructions must be carried out without fail. Otherwise undesirable concentrations of hydrocarbons may build up in plant vessels or in the drain lines.

2.2.2.1.3 Use of hydrocarbon absorbers. It is important that the specified operating period of an adsorber vessel is not exceeded, and that the required reactivation gas flow rate and outlet temperature are achieved.

Whenever the plant is restarted after a temporary stoppage of more than 1 hour, it is preferable to use the standby adsorber, provided that it has been fully reactivated. If the other adsorber is put back on stream, particularly if it is approaching the end of its cycle time, there is a danger that the warming during the stoppage may result in the desorption of contaminants into the distillation process.

2.2.2.1.4 Liquid level control. Where vessels are provided for scrubbing oil or hydrocarbons by means of liquid air, the level indicators on such vessels, e.g. hydrocarbon scrubbers, equalizers, should be operating properly and the correct liquid levels maintained.

2.2.2.1.5 Use of catalytic filters. On some air separation plants catalytic filters are fitted between the air compressor and aftercooler to oxidize hydrocarbon impurities present in the air feed to the plant. These filters are sometimes fitted with preheaters.

The type and amount of hydrocarbons oxidized depend upon the temperature of the catalytic filter. The plant manufacturer's instructions should be consulted, and the specified operating temperatures maintained.

2.2.2.1.6 The use of molecular sieve adsorbers. Modern plants are normally fitted with molecular sieve adsorption units to remove water and carbon dioxide from the air feed to the plant. A secondary function of a molecular sieve adsorption unit is to remove certain hydrocarbon impurities present in the air.

It is essential that the operating cycle times, regeneration temperatures and flows, and the temperature of the air entering the units, are maintained within the limits specified by the makers. If the plant is stopped for any reason, the outlet of the sieve unit should be shut off to prevent desorbed contaminants passing forward into the plant.

2.2.2.1.7 Reboiler–condenser operation. The reboiler or condenser in the low-pressure column of an air separation plant is a heat exchanger which condenses air or nitrogen vapours against constantly boiling liquid oxygen.

Plugging of oxygen passages with foreign material, including carbon dioxide or ice, can, under some conditions, cause hazardous concentrations of hydrocarbons to build up while the plant is operating at normal liquid levels.

A decrease in the reboiler liquid level below a certain point has the potential to concentrate hydrocarbon contaminants to a hazardous level. Therefore it is important to operate the reboiler at the full submergence level to maintain the boiling surfaces fully wetted. Changes in plant operating requirements may cause the reboiler liquid level to vary temporarily. During these changes, variation in level can be tolerated for short periods of time without affecting safe operation of the unit.

Acetylene, because of its low solubility in liquid oxygen (approx 9 ppm), presents a particular problem to the plant. When its concentration exceeds the solubility limit, solid acetylene can collect on the surface of the liquid or on the walls of the reboiler passages, where it creates high local concentrations capable of detonating. The presence of acetylene in liquid oxygen, when detected, requires immediate investigation and corrective action as recommended in the plant operating manual.

In plants with thermo-siphon reboilers, particular attention must be given to the reboiler liquid level when the air separation plant is on cold shutdown. The plant liquid should be disposed of whenever the liquid level falls appreciably or when the total hydrocarbon content exceeds the limit recommended by the supplier. Acetylene analysis of plant liquid must be conducted more frequently and plant liquid disposed of if the maximum permitted concentration is exceeded.

Liquid of acceptable quality may be transferred to the low-pressure column from storage, if available, to maintain a satisfactory liquid level and dilute the hydrocarbon concentration.

When starting a plant containing liquid after a cold shutdown, take special care not to boil away existing plant liquid or to permit desorbed hydrocarbons from the hydrocarbon adsorber to enter the low-pressure column. The plant operating manual should be consulted for the correct start-up procedures.

2.2.2.1.8 Monitoring of contaminants. Most hydrocarbon gases entering the plant are removed by freezing in the exchangers and by adsorption in the hydrocarbon adsorbers, except for methane and ethane, which are not readily adsorbed. Those hydrocarbons which can enter the low-pressure column are readily soluble in liquid oxygen, except for acetylene which is unstable and extremely hazardous. Acetylene is preferentially removed in the adsorbers.

A gas chromatograph or continuous hydrocarbon analyser can be used for monitoring the parts-per-million level of hydrocarbons within the oxygen-rich sections of the plant. The analyser selected for this duty will indicate when conditions change, and alert the operator to take corrective action should a potentially hazardous situation develop. The analyser should normally be used on a continuous basis to monitor a representative sample of liquid oxygen from the column sump, because hydrocarbons tend to concentrate at this part of the plant. However, samples of the air-feed stream and the liquid on the downstream side of the hydrocarbon adsorbers should be checked periodically to give early warning of increasing hydrocarbon contamination, especially when atmosphere pollution levels are known to be high.

Acetylene testing apparatus should be used daily to check the liquid oxygen in the low-pressure column, and weekly to check on the liquid oxygen in storage, or more often if required. These checks are particularly important if an acetylene plant is located in the vicinity of the air separation unit. To minimize the potential hazard from this source of contamination, acetylene plants should preferably be installed on the predominantly downwind side, and at a distance not less than 100m from the air separation plant air intake.

There will normally be a certain level of contaminants dissolved in the column sump liquid, but, provided the level is within the specified limit, the plant will operate safely under these conditions. Any significant increase above the normal background level will warrant an immediate investigation to trace the cause of the problem and take appropriate corrective action. If the corrective action (draining liquids, hydrocarbon adsorber changes, etc.) does not reduce the contaminant level, and the level of acetylene or total hydrocarbon contamination reaches the maximum recommended by the plant manufacturer, the plant should be shut down and the liquid in the columns drained off to the disposal system.

2.2.2.1.9 Plant de-rime (defrost) schedules. Trace contamination in the incoming air is removed from the process stream in the heat exchangers and in the hydrocarbon adsorbers. Most of the contaminants collected in the heat exchanger are removed when the reversing exchangers switch their cycles. Deviations from normal temperatures and pressures indicate a build-up of contaminants and show that defrost is necessary. An increasing contaminant level in the column and plant liquids indicates the need for hydrocarbon adsorber reactivation, defrost, or desiccant maintenance.

2.2.2.1.10 Draining of plant liquid after shutdown. When the plant is shut down for an extended period, i.e. over 24 hours, all liquid in vessels and process lines, except the main condenser, should be drained. The main condenser level should not normally decrease by much over a few days, but if it falls to half the normal operating level, the draining and disposal of all liquid will be necessary.

2.2.2.1.11 Decanting liquid oxygen or liquid nitrogen into an air separation unit. If liquid oxygen is decanted into an air separation unit, either to provide extra refrigeration during cooling down in normal operation, or to dilute hydrocarbon contaminated condenser liquid, it must first be tested to ensure that it contains less than the maximum amount of acetylene permitted for normal plant operation.

If liquid nitrogen is available, its use is preferable to that of liquid oxygen.

Under no circumstances should liquid be introduced if there is a known liquid leakage in the air separation unit.

2.2.2.2 *Contaminants introduced by equipment*

Lubricants from the plant's machinery (compressors or expanders) are possible contaminants of the process stream. On high pressure plants equipped with oil-lubricated reciprocating air compressors or expanders, lubricant carry-over from the cylinders is removed in the compressor entrainment separators, air driers, or adsorbers. However, periodic solvent washing of the plant may be necessary to remove trace quantities of lubricant carried over into the cryogenic portions of the plant.

For reciprocating machinery with dry-lubricated cylinders and oil-lubricated crank cases, it is important to maintain crank case, cylinders, oil slingers, piston-rod wipers, and similar devices in good condition to prevent the migration of oil from the crank case into the process stream via the rods and cylinders.

Low-pressure plants are equipped with centrifugal compressors and expanders which are designed to prevent lubricants entering the process gas by a more positive separation of oil-lubricated bearings from the process stream using combinations of distance pieces, seals, labyrinths, vents and purges. Purge, vent and seal gas systems must be operational on such equipment when the lubricating system is operating, to prevent contamination of the process gas stream.

2.2.2.3 *Contaminants introduced during maintenance and construction*

Maintenance work that requires opening the process system may accidentally permit entry of contaminants or unsuitable materials. Particular care must be taken during maintenance so that only proper materials and parts are used and that all new systems, parts and components have been appropriately cleaned. A system of inspection and checks should be used to avoid errors.

Parts, assemblies, accessories, and materials must be stored carefully to prevent contamination by dirt, mud, grease, oil, weld spatter, or as a result of the normal work environment or by bad weather. Material already prepared for oxygen service must be sealed and protected to avoid contamination during storage, and should be tagged and marked 'Clean for oxygen service'.

Personnel engaged in maintenance of plant machinery should wear clean overalls, preferably with zip fasteners and no external pockets.

2.2.3 Cold box purge and leak detection

The insulation spaces of air separation plant cold boxes are normally purged on a continuous basis with dry gaseous nitrogen. The prime purpose of the purge gas is to keep the insulating material dry by preventing ingress of water vapour from the atmosphere, but it also serves to prevent condensation of air on the external surfaces of equipment operating at liquid nitrogen temperatures. The presence of excessive quantities of ice in the cold box not only diminishes the thermal properties of the insulation but can also result in increased stress on internal piping and fittings.

The purge gas flow should be maintained at the specified rate, except when it is necessary to enter the cold box for repairs or when the plant has been shut down, drained and defrosted.

From time to time it is possible that leaks, either of gas or cold liquid, can develop within the cold box. Minor leaks may go undetected but those of an appreciable nature will become apparent in one or all of the following ways:

(a) Formation of white frost spots on the external surface of the cold box.
(b) An increase in pressure of the gas in the cold box insulation space.
(c) If the leak is oxygen, an increase in the oxygen content of the purge gas in the insulation space.
(d) In the event of a severe liquid leak, the cracking of mild steel panels and presence of vapour fog and liquid outside the cold box local to the area of leakage.

Regular monitoring of the cold-box surfaces and the pressure and oxygen concentration of the gas in the cold-box insulation space is recommended. This will increase the chance of detecting the development of a leak as early as possible. Naked lights, cigarettes, or similar heated sources must not be used to detect any suspected leaks on oxygen producing or handling plants. Gas analysis equipment will generally be the first means of confirming that an oxygen leak is present. Leaks on external pipe joints which are not obvious to the hand, eye or ear may be located by applying a suitable liquid detergent to the suspected area.

2.2.3.1 *Notification of leaks*

When there is evidence of enriched air due to oxygen or liquid air leakage in an air separation unit or from external oxygen lines, the supervisor in charge of the plant must be notified immediately so that he may decide the appropriate action to be taken.

2.2.3.2 *Operation of plant with liquid leakage*

If a gas-producing air separation unit has not sufficient refrigeration capacity to maintain correct reboiler operating level, and this is known or suspected to be due to internal liquid leakage, the plant must be shut down to rectify the fault. To maintain the plant in operation under such conditions by feeding in liquid from an external source is extremely dangerous.

2.2.3.3 *Areas affected by leaks*

Areas affected by leaks must be regarded as danger areas and the necessary precautions must be taken. Leaks of cold vapours or gases which are heavier than air, such as oxygen or argon, may accumulate in pits or trenches. If entry to these areas is contemplated, the atmosphere should be checked beforehand for oxygen content.

Large liquid leaks may cause frost heave, which could endanger foundations, etc.

2.2.4 Disposal of cryogenic liquids

The presence in the plant of liquid oxygen which has been heavily contaminated with hydrocarbons represents a potential hazard to equipment and personnel. Therefore quick action is required to dispose of such liquids to reduce the hazard. Collecting the liquid in a vessel outside the air plant does not eliminate the hazard, because the vaporizing liquid in the collecting vessel will also concentrate the contaminants.

Oxygen, air or nitrogen in liquid form must never be drained into trenches, pits, road drains, confined spaces or near personnel. The disposal system must vaporize the liquid without allowing any contaminants to accumulate. The heat supplied to the disposal vaporizer may be from water, steam, or ambient air. The vaporized liquid should be vented in a manner that will not expose personnel, work areas or equipment to increasing oxygen concentrations, which can contribute to the potential for fire; or to decreasing oxygen concentrations, which may give rise to an asphyxiation hazard.

Vents should be directed away from structures liable to mechanical failure on exposure to low temperatures.

2.2.5 Vapour fog clouds

Vapour fog clouds can form during certain phases of plant operation, such as draining hydrocarbon adsorbers of liquid before reactivation or transferring product from storage to liquid transport vehicles. The vapour fog is composed of atmospheric water vapour condensed by the cooling effect of the liquid being vaporized. The fog should also be assumed to contain a possible hazardous concentration of the vaporized liquid. Depending upon the quantity of product being vaporized, and climatic conditions, the fog can travel considerable distances, causing visibility problems for plant operators. Vehicle accidents may also occur if vapour clouds are blown across nearby public highways. Tests and calculations show that the extremities of the visible cloud are non-hazardous. If liquid oxygen is released into the atmosphere, the possibility of fire exists. The dispersion of oxygen vapour clouds can be calculated by referring to the British Compressed Gases Association document 'A method for estimating the off-site risks for Bulk storage of liquefied oxygen (LOX)'. If the product is inert, then asphyxiation can be a potential hazard.

Elevated vents from process equipment and control of the venting rate can reduce exposure to vapour fog clouds. Fans can inject air into the vents to dilute and heat the vapour and to enhance dispersal into the atmosphere. In this way the hazard to operating personnel or the public is minimized.

Ambient air vaporizers, when used for plant liquid disposal or as product vaporizers, can generate vapour clouds from the contact of air moisture with the cold metal portion of the vaporizer. The fog is not so dense as that caused by liquid venting direct to the atmosphere, but under certain atmospheric conditions the fog may collect in operating or public areas adjacent to the plant, and reduce visibility.

2.2.6 Thermal expansion

A hazard exists in a process system whenever cryogenic liquid can be trapped between closed valves. As the liquid vaporizes through normal heat leak, the pressure generated may cause the trapped section of piping or equipment to fail. Pressure relief valves, not designed to handle full process flows, are provided in these sections of the system to relieve the pressure generated by trapped liquid.

2.2.7 Modifications to plant

Proposed modifications to an air separation plant should be thoroughly reviewed to determine whether they affect the safety of the system or introduce new hazards. Modifications undertaken without the advice and experience of competent cryogenic engineers and operators can unknowingly introduce hazards, through improper choice or use of materials for the conditions of service, or through introduction of contaminants. Proposed changes in operating procedures should also be reviewed in the same way.

2.2.8 Special equipment hazards

The mechanical equipment for handling the oxygen and nitrogen products of an air separation plant and the transmission system for the gaseous oxygen product require special consideration because of oxygen's ability to support combustion. Such equipment contains recognized ignition energy sources, and it is therefore essential that clean components are used to eliminate the presence of easily ignitable fuels. Oxygen compressors, oxygen pumps, nitrogen compressors, and oxygen piping systems are examples of equipment containing recognized ignition sources.

2.2.8.1 Centrifugal oxygen compressors

Within a centrifugal oxygen compressor energy is being introduced to increase the pressure of the product, and components with very small clearances are moving at high relative speeds. Balance and stability of rotating parts are essential. Instabilities may be caused by surging, passing through critical speeds, and/or sudden unexpected shutdowns. Even momentary rubbing or impact between rotating and stationary elements may create temperatures sufficiently high to ignite the metal in the oxygen atmosphere and cause damage to the compressor.

Bearing wear occurring during normal operation or as a result of lubricating system failure is a condition which can lead to instability of the rotating elements of the oxygen compressor.

Failure of the seal gas supply to the labyrinth seals of the compressor increases the possibility of lubricating oil migrating into the process stream. Oil is a recognized low ignition temperature fuel which is capable of reacting with oxygen.

The most critical periods of instability and also the periods of greatest potential hazard exist during start-up and shutdown of the compressor. The following design and operating features should be considered and evaluated to reduce the exposure of personnel and equipment to the potential fire hazard associated with a high-pressure centrifugal oxygen compressor:

(a) Vibration sensors to detect axial and radial movement which alarm and shut down the compressor.
(b) Interstage high-temperature gas alarm and shutdown instrumentation.
(c) Seal gas pressure with alarm and shutdown points.
(d) Lubrication system pressure and temperature control system with alarm and shutdown points and an auxiliary oil pump to supply compressor bearing lubrication when the main lubricating oil pump is not operating.
(e) A purge system for start-up and shutdown, using clean, dry, oil-free nitrogen or air instead of oxygen.
(f) Location of the compressor behind a non-combustible barrier for personnel protection. Low-pressure oxygen blowers may require shielding only at critical areas of the machine.
(g) Location of the compressor where the potential fire exposure to other plant activities is minimized.
(h) Location of the control panel and critical instrumentation for the compressor outside the non-combustible barrier.
(i) Filters/screens in the compressor gas inlet lines to prevent particles from entering.
(j) Mandatory checklists must be completed before start-up.
(k) Compressor should preferably be started up on nitrogen before oxygen is introduced.
(l) All operating characteristics must be normal after start-up before oxygen is put through the compressor.
(m) All personnel must be outside the barrier before switching to oxygen and while compressing oxygen.
(n) Compressor maintenance must be rigidly controlled.
(o) Logged data to be periodically reviewed to detect trends.

2.2.8.2 Reciprocating oxygen compressors

The design requirements and operating procedures applicable to high-pressure centrifugal compressors apply also to high-pressure reciprocating compressors, except for the vibration, seal gas and surge control requirements.

Attention must be given to prevent oil migration along the piston rods. Slingers, distance pieces, and crankcase eductors are some of the devices used to control lubricant migration. In addition, a permanently mounted ultraviolet inspection light may be used to detect, by fluorescence, the presence of lubricating oil on the piston rods. A fluorescent additive can be added to the lubricating oil to improve ease of detection. If lubricating oil migrates as far as the cylinder end of the piston rods, a potentially hazardous situation exists, and the compressor must be shut down to investigate and correct the fault.

Compressor valve cover temperatures may also be monitored to detect changes in valve performance.

Complete enclosure of the compressor by personnel barriers may not be necessary with some applications.

2.2.8.3 Liquid oxygen pumps

Both centrifugal and reciprocating type pumps are used for handling and transferring liquid oxygen. Although the materials of construction of liquid oxygen pumps are carefully chosen, particularly those of the stationary and moving parts in contact with the fluid, a potential fire hazard can exist in the event of an abnormal operating condition. Insufficient cooling of the pump, cavitation or mechanical failure of shafts, pistons or bearings may create sufficient energy in the form of frictional heat to ignite or melt moving and stationary components in contact with oxygen.

Pump installations, particularly those having high speed or high head pumps, should incorporate features designed to protect the equipment and minimize the potential hazard. The following important safety measures should be considered for inclusion, as appropriate, in a liquid oxygen pump installation:

(a) Personnel protection barriers for high speed or high head pumps which are not located within a cold box or below ground level.
(b) Suction line filters, the mesh sizes of which are governed by the clearances between the moving and stationary parts.
(c) Safety devices to shut down the pump in the event of cavitation or loss of prime.
(d) The location of the control station at a reasonable distance away from the pump (3-5m).
(e) Provision of automatic devices to shut off the liquid flow to the pump in the event of pump failure.
(f) Use of ignition resistant materials, e.g. tin bronze, for the pump casing and impeller.

2.2.8.4 Nitrogen compressors

Nitrogen compressors present a unique problem. Fires have occurred in reciprocating oil-lubricated compressors pumping high purity nitrogen when the oxygen content has suddenly increased due to plant upset conditions.

Ignition can occur in centrifugal compressors pumping nitrogen through interference of rotating parts, and in reciprocating compressors through ring or valve failure, but only in the presence of sufficient oxygen. Fuel for the reaction can be lubricating oil or accumulated organic material which has normally been exposed only to an inert atmosphere.

Continuous analysis of oxygen in the nitrogen stream, coupled with alarms and shutdown switches, provide the necessary operating information and protection to reduce the risk of an ignition due to increasing oxygen content.

2.2.8.5 Pipelines

High velocity flow of oxygen gas in a pipeline is recognized as a potential hazard. Any dirt, metal scale, or other particle travelling with the high velocity flow represents a potential energy source (impact) capable of starting a reaction.

Several historical observations which have been made from studies of fires in oxygen service are listed:

(a) No ignition has been known to occur in oxygen-clean carbon steel or stainless steel systems sized for limited gas velocities unless initiated by the presence of a foreign substance.
(b) Ignition has occurred several times in carbon steel and stainless steel systems operating at, or near, sonic velocity. Friction from high-velocity particles is considered to be the most probable cause of ignition.
(c) Copper, brass and nickel alloys have a low heat of combustion relative to steels and iron, and are resistant to ignition in oxygen service. Once ignited, these materials burn at a much slower rate than carbon or stainless steels.
(d) Aluminium is resistant to ignition, although under certain conditions it can ignite in the presence of oxygen and rust from carbon steel, resulting in rapid combustion. The use of aluminium for gaseous oxygen pipelines is therefore avoided.

The velocity of oxygen in steel pipelines is limited to 200ft/s (60m/s) at pressures up to about 14 bar. Normally, velocities are considerably lower, to limit pressure drop and power consumption.

At valves, orifices, and line sections where the velocity can exceed 200ft/s (60m/s), copper, brass, or nickel alloys should be used. Monel tees or impact plates should be used at points of abrupt change in flow direction.

In general, oxygen gas pipelines should be kept free of contaminants and all valves in the system should be operated slowly to avoid sudden changes in flow and pressure.

2.2.8.6 Liquid storage containers

Product liquids are stored and transported in purpose designed and constructed containers of capacities ranging from a few tons to several thousand tons. Containers of capacities up to about 50 tons are normally of the vacuum insulated type operating at pressures in the region of 16 bar. Larger capacity tanks are also double-skinned, use conventional thermal insulating media and operate at lower pressures in the region of 0.5 bar.

Normally, cryogenic liquid storage tanks are installed and located in accordance with established codes of practice, which include measures for minimizing the effects of liquid spillage. Nevertheless, sudden and uncontrolled releases of large quantities of cryogenic liquid to the atmosphere are undesirable events which pose significant potential hazards, and every endeavour should be made to ensure that they are prevented.

Apart from basic design and manufacturing faults on the storage tanks, which are extremely rare, the principal causes of uncontrolled liquid spills are the following.

2.2.8.6.1 Overfilling. This may occur during transfer of liquid to a tank, either from a production plant or from a road tanker. Overfilling can cause discharge of liquid from vent lines normally discharging vapour, and can also cause a rapid build-up of the tank pressure, resulting in the discharge of liquid to atmosphere by way of the vapour space pressure relief devices. In an extreme situation tank rupture could occur, owing to excessive hydrostatic pressure.

In general, overfilling can be avoided by training of personnel in correct operating procedures, proper positioning, marking and maintenance of liquid level indicators and, where necessary, the provision of high level alarm and trip systems.

2.2.8.6.2 Overpressure. The development of excessive pressure in the tank vapour space can only occur if for some reason the pressure safety valves either fail to operate at their set pressures or their capacity is inadequate to handle the upset condition. Although pressure safety devices should be maintained and tested at regular intervals to ensure as far as possible that they are in working order, it is desirable to reduce to a minimum the number of demands upon them to operate.

The situations which usually give rise to an increase in pressure in cryogenic storage tanks are as follows:

(a) Heat leak. There will normally be a steady flow of heat into the tank which will determine the liquid boil-off rate. However, the rate of heat input may rise, owing to deterioration of the thermal insulation, or, in the case of vacuum insulated tanks, failure of the vacuum. The latter situation can occur suddenly, resulting in a rapid increase in the rate of heat leak, with subsequent increase in the liquid boil-off rate and rise in pressure.
(b) Overfilling as discussed above.
(c) Flash vapour due to the injection of relatively warm liquid into the tank vapour space. The liquid may be from a production plant, a road tanker or recycled from a liquid pump. Excessive liquid injection rates will clearly result in increased flash vapour and a tendency to raise the tank pressure.

Most vacuum-insulated tanks are equipped with a liquid fill connection at the bottom of the tank as well as at the top. The bottom fill connection can be used when filling from a road tanker to control the rate of rise in pressure in the tank vapour space.
(d) Failure of the pressure build-up valve in the open position. Most cryogenic liquid tanks are fitted with an automatically controlled pressure build-up system to maintain the pressure in the vapour space during periods of high liquid withdrawal rates. Failure of the automatic valve in the open position will clearly result in a continuing build-up of pressure in the tank until corrective action is taken or the pressure safety valve operates.

2.2.8.6.3 Underpressure. Low-pressure tanks are vulnerable to damage by the development of partial vacuum conditions in the vapour space, owing to the nature of the tank construction and the pressure at which they normally operate. Partial vacuum conditions can arise because of excessive liquid withdrawal rates or, under certain conditions, sudden increases in atmospheric pressure. In general the creation of partial vacuum conditions can be avoided by adherence to correct operating procedures, ensuring that the pressure build-up system is operating correctly and that the vacuum relief valve is in working order.

It must be recognized that frequent demands on the vacuum relief valve may give rise to ingress of excessive quantities of moist air, resulting in accumulation of ice and subsequent blockage of the relief valve piping.

2.2.8.6.4 Failure of ancillary equipment. Failure of valves, piping and fittings on liquid off-take lines from the base of liquid storage tanks can clearly result in large uncontrolled spillages of liquid. Liquid lines on larger capacity tanks may be provided with internal or external emergency shut-off valves to minimize the extent of liquid spillage in the event of such failures. However, demands on emergency devices should be minimized by ensuring that liquid lines, valves and fittings external to the tank are properly inspected and maintained and, as far as is practicable, protected from mechanical damage.

2.2.8.6.5 Failure of purge gas flow to insulation space. The insulation spaces of cryogenic liquid storage tanks of the non-vacuum insulated type are normally purged with dry gaseous nitrogen in the same way as the cold box of an air separation plant (see section 2.2.3).

Failure to maintain adequate flow of purge gas may result in ingress of air, causing accumulation of ice and, in the case of low pressure liquid nitrogen storage tanks, condensation of air on the external surfaces of the inner vessel. In extreme cases these conditions may adversely affect the integrity of the tank and internal piping.

2.2.8.6.6 Tow-away. Incidents have occurred when road tankers containing cryogenic liquids have been driven away from the tank fill point with the flexible transfer hose still connected. Such incidents have led not only to damage to piping and equipment but also to large spillages of liquid.

Although tow-away incidents can largely be prevented by proper training of drivers and other personnel responsible for liquid transfer operations, the experience of several companies has shown that this is not always sufficient, and that the potential consequences of a tow-away warrant the introduction of further safeguards. Such safeguards include flag warnings and various interlock systems which give audible warnings or prevent release of the vehicle brakes until the transfer hose is disconnected.

2.2.8.7 Road tankers

Road tankers used for the transport of cryogenic liquid should be maintained and operated in accordance with the manufacturers' instructions and any relevant statutory requirements.

Liquid transfer hoses should be capped when not in use to prevent ingress of water and dirt. Hose couplings used for the transfer of a particular cryogenic liquid should be non-interchangeable with those used for other liquids.

The practice of changing the service of a road tanker from one cryogenic liquid to another on a frequent basis should be avoided. Such situations, unless carefully controlled, can lead to problems with product contamination, particularly where it is necessary to use adaptors for connecting hose couplings normally used for one liquid to fill points designed for another.

Strict controls are required when tankers in inert gas service are converted to liquid oxygen service.

It is essential to ensure that drivers and other personnel engaged in road tanker and liquid transfer operations receive formal education and training in the potential hazards and the correct operating procedures.

2.3 Safety in the maintenance of air separation plants

The establishing of proper maintenance programmes and practices is of vital importance to ensure the continuing safe operation of an air separation plant. However, the unique features of air separation plants and the nature of the gaseous and liquid products require that certain special safety precautions are taken, both in the workshop and when carrying out maintenance operations on plant equipment *in situ*.

The purpose of this section of the guide is to identify those key areas of safety which are recognized by the industry as being of particular importance for the successful implementation of a maintenance programme.

2.3.1 Equipment

The main items of equipment which will require regular attention to maintain them in good order are:

(a) Machinery, e.g. compressors, pumps.
(b) Vessels, piping and valves.
(c) Insulation.
(d) Instruments.
(e) Safety devices.
(f) Yard areas.
(g) Auxiliary services (water, steam, power, instrument air).

In view of the high degree of reliability which is required of continuous process plants it is essential that a well-planned preventive maintenance programme is prepared and implemented. Regular attention to key areas is vital to minimize the occurrence of accidents or major breakdowns, which can be costly in terms of injury, repairs and business interruption.

2.3.2 Cleanliness

It has already been stated that oxygen vigorously supports combustion and, under the right conditions, can react violently with certain materials, particularly oils and greases. For this reason cleanliness is of critical importance for all equipment in service with oxygen or oxygen-enriched fluids.

It is therefore necessary to start with a clean plant and to maintain the same standard of cleanliness by periodic cleaning and decontaminating procedures. Initially, all materials of construction, sub-assemblies and individual components comprising the air separation plant and product storage equipment will be subjected to cleaning processes. Once clean, all equipment and components should be protected during construction and thereafter protected or recleaned in accordance with established repair or maintenance procedures.

2.3.2.1 Maintenance facilities

2.3.2.1.1 Special maintenance area. An area for final cleaning and assembly of air separation plant components and other equipment in service with oxygen must be clearly defined, segregated from other maintenance areas, and kept in a clean, oil-free condition. The use and storage of combustible materials in this area should be kept to a practicable minimum. Dismantling and initial cleaning of equipment and components must be carried out before they are taken to the clean area for repair.

For maintenance of special items of equipment, such as liquid oxygen pumps, which require the highest standards of cleanliness, a special clean room should be provided to minimize the likelihood of contamination during repair and reassembly.

2.3.2.1.2 Work benches. Work benches in the special maintenance areas must be covered with stainless steel or other non-combustible, non-absorbent material, and kept clean and free of oil and grease. Relatively soft materials such as wood or aluminium, are not recommended.

2.3.2.1.3 Tools. Tools for maintaining and assembling air separation plant components must be kept thoroughly clean and used only for this purpose. Tools that are used for general maintenance work, such as crankcase overhaul, must not be used for maintenance or assembly of air separation plant components or equipment which have been cleaned for oxygen service.

2.3.2.1.4 Protective clothing. Protective clothing, such as overalls, footwear, gloves, etc., used when carrying out maintenance operations on air separation plant components and equipment, must be clean and free from oil and grease. This is of prime importance when carrying out work on special items of equipment such as oxygen turbocompressors. Overalls without external buttons or pockets are best, to minimize the chances of foreign objects such as pens, buttons, etc., entering clean equipment.

2.3.2.1.5 Personal cleanliness. Personnel must cleanse hands thoroughly and change into approved protective clothing before transferring from general maintenance work involving dirt, oil, or grease, to air separation plant maintenance. Barrier protection creams usually contain combustible components, and must not be applied to the hands before handling components which have already been cleaned for air separation plant service.

2.3.2.2 Standards of cleanliness

Guidance on the frequency and methods of cleaning and on the degree of cleanliness required should be obtained from the plant manufacturer, and the

Oxygen, nitrogen and argon 47

instructions should be carefully followed. In general, all equipment should be free of hydrocarbon oils and greases, loose or potentially loose slag, scale, metallic chips, solvent residue and other foreign material before use in oxygen service.

2.3.2.3 Inspection techniques

Various techniques are used for inspecting equipment and components to determine whether an acceptable standard of cleanliness has been achieved. Such methods include:

(a) Bright white light to inspect surfaces for particles, fibres and liquid films.
(b) Ultraviolet light ('black light') to detect fluorescent oil or grease deposits.
(c) Wipe tests with clean white filter paper for detecting residues not detectable by ultraviolet light.
(d) Solvent extraction test to check inaccessible surfaces by washing and subsequent quantitative analysis for contaminants.

The criteria for the required degree of cleanliness of equipment for oxygen service, as determined by the above inspection methods, should be obtained from the plant manufacturer.

2.3.2.4 Cleaning methods

2.3.2.4.1 Method selection criteria. Selection of the most practical and efficient methods for cleaning equipment and components for oxygen service will depend on:

(a) The size of the item of equipment.
(b) The mechanical properties of the equipment and its component parts, e.g. temperature effects.
(c) The location and degree of contamination.
(d) The nature of the contaminants.
(e) The arrangement of passages with respect to their ability to be flushed and drained.
(f) The compatibility of the cleaning agent with the contaminants and the materials of which the equipment is constructed.
(g) The availability and cost of cleaning methods plus the availability of cleaning agents and personnel trained in their handling and use.
(h) The speed and effectiveness of the cleaning method and its ability to meet the desired level of cleanliness.

In general, cleaning of equipment may be accomplished by any method which satisfies the acceptance criteria as recommended by the plant manufacturer. Any solvent or chemical agent used in the cleaning process should be a commercially approved type of satisfactory grade and quality and compatible with the materials of construction of the equipment being cleaned. However, the use of combustible fluids, such as paraffin, acetone, or petrol, is not recommended for cleaning equipment for oxygen service.

2.3.2.4.2 Cleaning processes. The principal methods used for cleaning process equipment include:

(a) *Steam or hot water cleaning.* Steam or hot water is applied through a nozzle or spray, usually in conjunction with a detergent, for removal of contaminants such as dirt, oil, and loose scale. The steam and hot water should be oil-free; if they are not a secondary cleaning operation with a solvent may be necessary (see section (e) below).

(b) *Mechanical cleaning*. This type of cleaning may be accomplished by shot blasting, wire brushing or grinding for removal of mill scale, rust, slag, grit, varnish, paint, or other foreign matter. Loose particles may be removed by vacuum cleaning after the mechanical cleaning process. Removal of metal and other particulate matter from oxygen systems, particularly machinery and carbon steel piping, is of prime importance.

(c) *Caustic cleaning*. Strong alkaline solutions are used to remove heavy or tenacious surface contaminants. Such solutions are generally more effective when warm, and may be applied by spraying, immersion or flushing.

(d) *Acid cleaning* This method may be used to remove oxides and other contaminants by immersion or flushing with a suitable acid solution, usually at room temperature. Commonly used acids are phosphoric acid or hydrochloric acid. The former is used on most metals for removal of oxides, light rust, and fluxes, whereas the latter is used primarily on carbon and low alloy steels for removal of rust, scale, and oxide coatings. Cleaning of carbon and low alloy steels by this method should include a final passivation wash to ensure complete neutralization of the acid. This is particularly important with brass components, where the presence of acids can initiate stress corrosion cracking.

(e) *Solvent washing*. This procedure is widely used for the removal of organic contaminants, such as oil or grease, from surfaces by the application of chlorinated hydrocarbons or other suitable solvents. Commonly used solvents include inhibited grades of 1,1,1 trichloroethane, and trichloroethylene, although the latter is used only in closed systems for vapour degreasing. The physical and chemical properties of solvents vary considerably, and have an important bearing on the selection of a solvent for a given cleaning method. The health hazards associated with solvents must be fully understood and assessed before use to ensure that all necessary precautions are taken. Reference should be made to the manufacturers' literature and the following relevant documents: HSE Guidance Note EH40, 'Occupational Exposure Limits', and 'The Control of Substances Hazardous to Health Regulations 1988: SI No. 1657'.

Equipment may be solvent-cleaned by immersion, flushing by recirculation or vapour degreasing. The efficiency of immersion cleaning may be augmented by the application of ultrasonic techniques, using a high-frequency sound generator. As for other cleaning processes, the choice of the cleaning method will depend on a number of factors. The main advantages of immersion or flushing techniques are that the cleaning equipment required is relatively simple. The disadvantages are that larger quantities of solvent may be required, and as the solvent becomes contaminated with foreign organic material it may require replacing with fresh solvent to achieve a satisfactory level of cleanliness.

Vapour degreasing is a process whereby soluble organic materials are removed from equipment surfaces by continuous condensation of solvent vapour and its resulting washing action. The main advantage of this method is that a high degree of cleanliness can be achieved, since the surfaces are always being washed by clean solvent and only a relatively small amount of solvent is necessary. The main disadvantage is that vapour degreasing equipment is complex and costly to install.

The environmental aspects of chlorinated solvents do not have the same ozone depletion effects as CFCs, but in the medium to long term these solvents should be replaced by environmentally acceptable alternatives.

2.3.2.4.3 Rinsing, drying and packaging. On completion of any cleaning process it is essential that all traces of chemical agents or solvents are completely removed from the equipment by draining and either rinsing with clean, oil-free water or purging with clean, dry, oil-free air or nitrogen as appropriate. The latter media may also be used to thoroughly dry equipment which has been rinsed with water.

It should be noted that chlorinated solvents can form flammable atmospheres in the presence of oxygen, and it is therefore of great importance to ensure that all traces of solvent are removed from the equipment or system after cleaning.

Once equipment has been inspected to ensure that the required standard of cleanliness has been achieved, it should be packaged or sealed in nitrogen to protect the surfaces from contamination until such time as it is put back into service.

2.3.2.4.4 Precautions. Practically all the cleaning methods described above call for the use of chemicals, solvents, or processes which are potentially hazardous. It is therefore essential that cleaning operations are carried out by properly trained and experienced personnel using suitable purpose-designed equipment. Personnel should be aware of the nature of the hazards, e.g. whether toxic or corrosive, and use appropriate protective equipment.

Cleaning agents should be used only if their performance and application are known, or after discussion with the cleaning agent manufacturers. The manufacturers' instructions for use should be strictly followed.

2.3.3 Precautions when handling perlite

Where maintenance or repair is required on a pipe, vessel or associated equipment which has contained a cryogenic liquid, insulating material must be removed before work can begin. For large-scale vessels and pipes in air separation unit cold boxes and low pressure cryogenic liquid storage tanks, an insulating medium called 'perlite' or 'brelite' is used. Perlite is an inert material made from expanded volcanic rock, which is pumped as very fine granules into the interspace of cold boxes or storage tanks. As perlite is inert, it is difficult to envisage any risks when handling it, but experience has shown that there are potential hazards associated with the material. Perlite can be discharged from the interspace of a cold box in the event of box failure and produce a cloud of particles which can reduce visibility, particularly on adjacent roads. Because it flows so easily, perlite can cause death by choking if ingested; this could occur if a person fell into a container filled with perlite.

2.3.4 Materials for maintenance

2.3.4.1 Use of approved materials

Air separation processes mostly call for the use of equipment in service with gases and liquids at elevated pressures, high and low temperatures, and, in parts of the plant, with oxygen. It is therefore vital when carrying out maintenance work to ensure that all replacement components and materials are of a type approved by the plant manufacturer and suitable for the conditions of service, namely:

(a) Pressure.
(b) Temperature.
(c) Flow conditions.
(d) Process fluid(s).

For parts of the plant in service with oxygen or oxygen-enriched fluids, components and other materials must be suitable for oxygen service in the particular application. The plant manufacturer will advise which materials are approved for use.

The use of incorrect components or materials can cause mechanical failures, fires, or uncontrolled release of gases or liquids, which may result in injuries to personnel or damage to property.

2.3.4.2 Types of material and component

In general, the types of material and component which must be considered when repair or replacement is contemplated can be categorized as follows:

(a) Piping, valves, and fittings.
(b) Jointing and gland materials.
(c) Jointing liquids and compounds.
(d) Thread sealants.
(e) Lubricating oils and greases.
(f) Other lubricating agents.
(g) Thermal insulating materials.
(h) Pressure gauges.
(j) Pressure safety valves.
(k) Safety instrumentation.

2.3.4.3 Storage and identification of spares

Spares must be stored under clean conditions in sealed packages or suitably protected against contamination or deterioration. The innermost packaging must be transparent and sealed to enable checking of the contents without exposing the spares to contamination or damage. Spares must be identified as conforming to a specification covering the composition, method of manufacture, and dimensions.

Users should ensure that all spare parts are supplied in accordance with an agreed specification. Problems can arise when manufacturers change manufacturing methods or specifications without warning.

2.3.5 Safety control procedures

Refer to section 1.2.9.

Bibliography

ASTM. 'Flammability and Sensitivity of Materials in Oxygen-Enriched Atmospheres'. ASTM Symposium Series, STP 812, STP 910, STP 986 and STP 1040.

British Compressed Gases Association. 'Code of Practice for Bulk Liquid Oxygen Storage at Production Sites' (CP20), 1990.

British Compressed Gases Association. 'Code of Practice for Bulk Liquid Oxygen Storage at Users' Premises' (CP19), 1990.

British Compressed Gases Association. 'A Method for estimating the Offsite Risks from Bulk Storage of Liquefied Oxygen' (LOX), 1984.

British Cryogenics Council. 'Hazards with Nitrogen and Argon Safety Package' (1984).

European Industrial Gases Association, Brussels. 'Bulk Liquid Nitrogen/Argon Storage at Production Sites'. Industrial Gases Council Doc. 25/85.

European Industrial Gases Association, Brussels. 'Bulk Liquid Oxygen Storage at Production Sites'. Industrial Gases Council Doc. 21/85.

European Industrial Gases Association, Brussels. 'Cleaning for Oxygen Service-Guidelines'. Industrial Gases Council Doc. 33/86.

European Industrial Gases Association, Brussels. 'Code of Practice for the Design and Operation of Centrifugal Liquid Oxygen Pumps'. Industrial Gases Council Doc. 11/82.

European Industrial Gases Association, Brussels. 'Fire Hazards of Oxygen and Oxygen Enriched Atmospheres'. Industrial Gases Council Doc. (to be published - replacement for Doc. 8/76).

European Industrial Gases Association, Brussels. 'Hazards of Inert Gases'. Industrial Gases Council Doc. 44/90.

European Industrial Gases Association, Brussels. 'Liquid Nitrogen and Liquid Argon. Storage Installations at Users' Premises'. Industrial Gases Council Doc. 17/85.

European Industrial Gases Association, Brussels. 'Liquid Oxygen Storage Installations at Users' Premises', Industrial Gases Council Doc. 16/85.

European Industrial Gases Association, Brussels. 'Reciprocating Compressors for Oxygen Service'. Industrial Gases Council Doc. 10/81.

European Industrial Gases Association, Brussels. 'The Transportation and Distribution of Oxygen by Pipeline'. Industrial Gases Council Doc. 13/82.

European Industrial Gases Association, Brussels. 'Turbo Compressors for Oxygen Service'. Industrial Gases Council Doc. 27/82.

HSE.'Fires and Explosions due to Misuse of Oxygen'. Health and Safety Executive, Leaflet, 8 (1984).

3 Natural gas, ethylene and ethane

Chapter 3 deals with special precautions which must be observed in the supervision, operation and maintenance of liquefaction, product handling and storage of natural gas, ethylene and ethane. The general term 'plant' is used in the following sections to define liquefaction, storage, boil-off control, and vaporization to generate gas for export from storage.

This chapter supplements the general safety requirements of chapter 1.

Table 3 gives some of the important properties of methane, ethane and ethylene.

3.1 Specific hazards

3.1.1 Fire and explosion hazard

3.1.1.1 General

Natural gas, ethylene and ethane in certain concentrations with air or oxygen form flammable mixtures; care must therefore be taken to prevent uncontrolled releases to atmosphere. The composition of natural gas is predominantly methane with approximately 5 per cent by volume of higher hydrocarbons. However, the composition will tend to vary depending on the source of the gas.

Because of the low temperatures at which the liquids are stored, any release is initially heavier than air and will tend to fall to ground level and spread until the gas temperature rises to a point at which the gases are lighter than air or neutrally buoyant. A liquid release will quickly boil off to gas, giving a large volume increase.

3.1.1.2 Ignition and burning characteristics

3.1.1.2.1 Spontaneous ignition temperature (SIT). Natural gas, ethylene and ethane have spontaneous ignition temperatures of about 540°C (813K), 450°C (723K) and 510°C (783K) respectively. These values depend on the purity of the gases and to some extent on the measurement conditions.

3.1.1.2.2 Laminar burning velocity and significance of confinement The laminar burning velocities of natural gas and ethane are low compared with

Table 3 Thermophysical properties of methane, ethane and ethylene

	Methane	Ethane	Ethylene
Chemical symbol	CH_4	C_2H_6	C_2H_4
Molecular weight	16	30	28
Normal boiling point °C (K)	−161 (111.7)	−89 (184.6)	−104 (169.3)
Freezing temperature °C (K)	−183 (90.6)	−183 (89.9)	−169 (104.2)
Critical temperature °C (K)	−82 (191)	+32 (305)	+10 (283)
Critical pressure, bar abs	47	49.7	52.5
Expansion ratio-increase in volume as liquid at 1 bar abs boils to gas at 1 bar, 15°C	626	437	489
Density of saturated liquid at 1 bar abs ($kg\,m^{-3}$)	424	546	565
Relative gas density (referenced to dry air at 1 bar abs, 15°C, density 1.21 $kg\,m^{-3}$)	0.56	1.05	0.97
Latent heat of vaporization (cooling, potential of phase change) $kJ\,kg^{-1}$	512.4	488.3	483.4
Air liquefaction hazard	No	No	No
Flammable limits in air (vol. %)	5–15	3–12.4	2.7–36
Spontaneous ignition temperature in air °C (K)	540 (813)	510 (783)	450 (723)
Minimum ignition energy (mJ)	0.28	0.24	0.085
Flame temperature °C (K)	1880 (2153)	1895 (2168)	1975 (2248)
Limiting oxygen index (vol.%)	11.5	11.0	11.5

other hydrocarbons (less than $0.5\,ms^{-1}$). However, ethylene has almost twice the velocity of natural gas. The burning velocity is an indication of the tendency of a flame to 'run up' to deflagration.

It has been shown that, given the appropriate degree of confinement and congestion, all these fuels can 'run up' to a deflagration. Detonation, although not impossible, is unlikely.

3.1.1.2.3 Minimum ignition energy. Spark ignition energies may be used as a measure of sensitivity to ignition by localized sources. Various sources quote slightly different values. However, natural gas, ethylene and ethane minimum values are quoted as 0.28, 0.085, and 0.24 millijoules respectively.

3.1.1.2.4 Flammable range. Ethylene has a much wider flammable range than natural gas or ethane. In air at atmospheric pressure ethylene has a flammable range of 2.7 to 3.6 per cent by volume whereas natural gas has a range of 4.9 to 15 per cent and ethane 3 to 12.4 per cent by volume.

3.1.1.2.5 Limiting oxygen index. The limiting oxygen index is the minimum concentration of oxygen to support flame propagation when a stoichiometric fuel–air mixture is diluted with nitrogen, that is, it is the least amount of oxygen required to support combustion. For methane, ethylene, and ethane the values are 11.5, 11.0 and 11.5 per cent by volume respectively.

3.1.1.2.6 Quenching distance. This is the minimum gap between two parallel surfaces which will just permit a flame to pass without cooling it to extinction. The smaller the quenching distance, the greater resistance the flame has to cooling, and the more difficult it will be to extinguish. The maximum experimental safe gaps for natural gas, ethylene, and ethane are 1.14mm, 0.65mm and 0.91mm respectively.

3.1.1.2.7 Flame characteristics. Non-aerated hydrocarbon flames have a yellow tinge. The flame temperatures of natural gas and ethane are similar (1880°C and 1895°C) but ethylene is higher at 1975°C.

3.1.1.2.8 Thermal expansion. Equipment and pipework containing a cryogenic liquid could be subject to severe overpressure, owing to thermal expansion of the liquid, in the event of liquid being trapped in a closed system. This will result from heat inleak or loss of secondary cooling.

Overpressure can also be generated by vaporization of a cryogenic liquid in a closed system. Pressure relief valves are provided in these sections of the system.

3.1.1.2.9 Decomposition. Ethylene or an ethylene/diatomic gas mixture can undergo thermal decomposition reactions at elevated pressures in certain circumstances. Care must be taken to avoid the rapid pressurization of systems over a large pressure range, and also to avoid 'hot spots' on equipment such as pumps, compressors, etc., which may initiate decomposition.

Examples of precautions taken are that on high pressure (HP) systems the system to be reconnected to the HP system is first pressurized with inert gas to the operating pressure before opening the isolation valve. Rotating equipment requires full thermal and mechanical monitoring to detect impending mechanical failure or temperature rises in order to keep the equipment within acceptable operating ranges.

3.1.2 Oxidants and their avoidance

The oxidizing agents which may exist in plant are air and cold-box atmosphere containing air diluted with nitrogen. It should be remembered that air will initially be present in cold-boxes, process vessels, and pipework.

3.1.2.1 Purging

The limiting oxygen index is relevant only to the problem of flammability. Considerations such as solubility, liquefaction, or freezing at the process temperature may require limiting the oxygen to much lower concentrations than those dictated by flammability alone.

Before the introduction of a hydrocarbon into processing equipment, air should be eliminated by nitrogen purge to give an exit oxygen concentration of no more than 2 per cent oxygen, i.e. approximately 20 per cent of the minimum oxygen index. To allow for sampling errors, a figure lower than this may be used in practice.

3.1.2.2 Air ingress

Operation below atmospheric pressure should be avoided unless the plant is designed for vacuum conditions, as this could lead to the ingress of air and the formation of flammable mixtures. Suitable safeguards in operating instructions or hardware should be provided to prevent operation below atmospheric pressure where applicable.

3.1.2.3 Condensation of air and other gases

Natural gas, ethylene, and ethane at their boiling points will not cause atmospheric oxygen to condense. At their melting points only natural gas is close to the boiling point of oxygen.

On the other hand, all these gases when liquefied can solidify carbon dioxide. Hence it is important to remove carbon dioxide from plant by purging, and from the gas to be liquefied by pre-treatment, otherwise blockages may occur in the process.

3.1.3 Electrostatic effects

Electrostatic charges can build up as a result of friction in pipes or by the break-up of liquid hydrogen to droplet size. This can cause ignition where gas is venting to atmosphere.

3.1.3.1 Safeguard against electrostatic build-up

Routine inspection to ensure the integrity of electrical earthing and bonding systems on plant equipment is essential. When maintenance work has been carried out, the electrical bonding should be checked before recommissioning.

3.1.3.2 Road tanker operations

Before transfer is begun, tankers carrying the subject cryogens must be equipped with an earth cable, which must be connected to the earth terminal of the receiving equipment and remain connected until transfer is completed.

3.1.4 Firefighting

Natural gas, ethane and ethylene fires need to be fought from a safe distance and from the upwind side of the fire.

Even relatively small hydrocarbon fires are difficult to extinguish, the only reliable approach being to shut off the supply. In fact, until the source can be shut off, it may well be inadvisable to attempt to extinguish the flame, as this could result in the formation of potentially explosive mixtures. The following points are important:

(a) Liquid hydrocarbon when exposed to the atmosphere will produce a cloud of ice/fog from the air. The flammable mixture will probably be inside the vapour cloud and it is advisable that personnel keep well outside the area of visible moisture.
(b) Use large quantities of water, preferably in the form of a spray, to cool adjacent equipment and any burning material below the ignition point. Do not apply directly to a pool of burning liquid hydrocarbon, as water will evaporate additional gas.

(c) Depending upon the circumstances, it is not usually advisable to extinguish a hydrocarbon flame in confined areas if the hydrocarbon supply cannot be shut off.

The continued escape of unburned hydrocarbon can create an explosive mixture which may be ignited by other burning material or hot surfaces. It is better to allow hydrocarbons to burn in localized areas and keep adjacent objects cool with water rather than risk the possibility of an explosion.

(d) If electrical equipment is in danger from the fire, disconnect the supply or use carbon dioxide or dry chemical extinguishers above, and not directly on to, the fire.

3.1.5 Secondary hazards

These take the following forms:
(a) Pressure rupture (thermal expansion), due to trapping of cold liquid or vapour.
(b) Brittle fracture, due to metals being used below their embrittlement temperatures.
(c) Ice formation – if water is present in plant and equipment before cooling down, fractures may occur, owing to the expansion of ice formed. Plant dryness is therefore vital.
(d) Contraction leakage – sudden cooling can lead to contraction and leakage from bolted flanges and pipework. The system will have been specially designed to overcome this problem, provided cooling down is carried out at a specified uniform rate.

3.1.6 Health hazards

The hazards to health of these hydrocarbons stem from their low temperatures in the liquid phase and the exclusion of oxygen arising from vaporization in enclosed spaces.

Serious tissue destruction similar to that caused by burns can occur when the liquid comes into direct contact with body surfaces, owing to the low temperature of the liquid. Similar effects can arise from body contact with uninsulated equipment containing cold fluids. The cold gases can cause damage to the lungs if inhaled.

These hydrocarbons are non-toxic, but oxygen exclusion will result in asphyxiation, and it should be noted that this can often occur so rapidly that the victim may be unable to escape even if he realizes that something is wrong.

3.2 Safety in operation

3.2.1 General

It is essential to maintain safe working conditions by minimizing the possibility of release to atmosphere of these hydrocarbons in liquid or gaseous form. Also, it is important to prevent oil, grease, aromatic hydrocarbons, carbon dioxide, or water from entering the cold parts of the plant where blockage due to freezing may occur and lead to subsequent equipment malfunction. Similar attention should be paid to the cleanliness of equipment before recommissioning. Compressor suction lines are particularly vulnerable areas, where special care is required to ensure mill scale and other debris are removed and cannot be carried forward to the compressor. Serious damage and plant malfunction could occur either in the main working parts or in the protective and lubrication systems, which are designed to operate with fine clearances.

3.2.2 Ventilation of plant buildings

The plants usually include three distinct types of building where specific safety considerations apply, i.e. control rooms, compressor houses, and analyser houses.

It is claimed, not without justification, that the safest plant building is one with 'no roof and no sides'. Buildings in which flammables are handled should be as open as possible to permit safe operation with good natural ventilation in order to dilute any potential leakage to a safe concentration. Care should be taken to ensure that natural ventilation, as allowed for in the design, is not reduced by temporary or other modifications. Where forced ventilation systems are installed, the equipment, associated alarms and protective devices should be regularly tested and maintained.

3.2.2.1 Control rooms

Generally, these are designed not to contain any flammable gases, but because of their location relative to the plant, the ventilation systems may be required to ensure safe conditions in the event of any atmospheric release of flammable gas. In the case where the control room is pressurized or swept to prevent entry of flammable gas, safe operating conditions should be maintained by ensuring that the ventilation equipment alarms are regularly tested and that immediate action is taken in the event of system failure.

3.2.2.2 Compressor houses

To achieve maximum safety, compressor house design will give a substantially open building, with side walls having a large free space on all sides and with substantial free area in the roof, e.g. a ridge ventilator.

Natural ventilation of buildings should not be restricted by temporary or permanent plant modifications when flammable gases/liquids are present. Where noise levels are a problem and acoustic enclosure of the compressors is necessary, adequate ventilation and noise control can be achieved simultaneously by the use of acoustic louvres.

3.2.2.3 Analyser houses

Closed analyser houses normally contain specialized and non-flameproof equipment. In the event of ventilation failure or build-up of flammable concentrations of gas, the equipment has to be automatically isolated to maintain safe conditions. Ventilation equipment and alarm trip systems should be regularly tested and maintained to ensure that safe conditions prevail in the event of loss of ventilation.

3.2.3 Road tanker operations

The loading and off-loading of tankers requires special care to ensure that safety in operation is maintained. While each installation will have individual details in design and operation methods, the main areas requiring close attention are given below.

(a) *Filling*. It is essential to avoid overfilling. The tanker should either be weighed before and after filling, or filled with a metered quantity of product, or both. The maximum filling level may be dictated either by the maximum tanker barrel capacity or by the total road weight of the tanker.
(b) *Product compatibility*. Each tanker should be checked for previous cargo to establish if it is safe to load, whether inerting is required, and that the product to be loaded is fully compatible with the tanker.
(c) *Overpressure*. The operating pressure of the tankers during the filling, venting and purging operations should be known, and the systems set up to ensure that overpressure is avoided by adequate relief valves as well as by the operating procedures.
(d) *Tanker connection and disconnection*. Before connection for loading, the tanker documentation procedures should have been cleared.
   Before loading, all safety interlocks, earthing, venting, etc., should be checked as operational; procedures to prevent the tanker being driven away should be completed; and the tanker and loading system should be checked for leaks.
   To prevent spillage in the event of hose breaks, the ends of the hose should be protected from forward or reverse flow by means of emergency isolation valves and non-return valves respectively. To minimize the pressure in the hose and avoid running the tanker engine, fixed pumps rather than a tanker pump should be used. The reverse interlock procedure should be carried out before the tanker is allowed to leave.

3.2.4 Road tankers/small storage installations

A typical small facility to supply liquid fuel gas (LFG) should include the following features:

(a) Concrete bundwall to contain the total contents of the storage tank (see section 3.4.1).
(b) Fixed, dry-powder, fire-extinguisher installation at the tanker loading bay.
(c) Catchment pit to contain a potential spillage and thus minimize evaporation area.
(d) Portable dry-powder extinguishers for fighting small LFG fires.

Figure 10 LNG storage tank and bund wall (courtesy of British Gas plc)

(e) Water hydrant points for the supply of water to cool adjacent equipment exposed to heat from a fire.
(f) Vehicle barrier to safeguard installation from impact by vehicles.

3.2.5 Vehicle entry into plant areas

3.2.5.1 Administrative precautions

This section covers some basic precautions regarding entry of non-routine and maintenance vehicles into safe plant areas. Entry into high risk areas is covered in section 3.2.5.2. In general, vehicles may be classified in three groups:

(a) Vehicles passing through for short periods, where adequate control can be achieved by administrative procedures and access restricted to safe areas only.
(b) Part-time vehicles, e.g. cranes, where in addition to administrative procedures, a degree of protection to the vehicle is required.
(c) Vehicles requiring entry for more than 1000 hours per year, e.g. a forklift truck, where thorough protection of the vehicle is required.

All vehicles should be checked before entry. Vehicles and their destinations on the plant site should be logged on entry and cancelled on exit in order that in the event of an emergency the operator is aware of the number and whereabouts of vehicles on the plant.

Drivers should be made aware of the route to and from the location in the plant to which they require access, and be issued with a road map if the plant has a complex system. If the vehicle requires access to or is taking a route through a high-risk area, the control room operator should be notified before the vehicle proceeds from the plant gate.

Before entry, the reason for entry should be established, the driver dematched, and a check made that the vehicle is carrying a fire extinguisher.

Once on the plant the vehicle should not be allowed to block the plant roads or be left unattended with the engine running.

3.2.5.2 Vehicle entry to high risk areas

A high risk area in this context is a Zone 1 or Zone 2 area as defined in the 'Institute of Petroleum Electrical Code' and by BS 5345, Part 1. Additional administrative procedures may be required to cover the need for issue of fire permits in certain circumstances, positioning of gas detectors, and suitable supervision and inspection of the vehicle such that at any time it may be shut down safely.

With regard to protection of the vehicle itself, where frequent entry is required (see section 3.2.5.1.(c)), reference to BS 5908 ('Code of Practice for Fire Precautions in the Chemical and Allied Industries') will give details of protection considered necessary. Briefly, the requirements cover air intake protection to avoid overrunning, spark arrestors and a positive shut-off, exhaust protection to avoid sparks and excessive temperatures, and electrical-equipment modifications to achieve the required safety level.

3.3 Safety in maintenance

3.3.1 Responsibility for maintenance

The factory manager is responsible for the safe condition of the plant, and should ensure that regular maintenance and testing of the plant equipment and safety systems are carried out by competent personnel. Any plant modifications should be checked for their safety implications and authorized by a competent authority before installation and commissioning. Vessels containing steam, air and other fluids are subject to periodic statutory inspections in accordance with the Pressure Systems and Transportable Containers Regulations 1989 (SI No. 2169).

Other pressure vessels and systems containing hydrocarbons require inspections at set frequencies which should be established before initial plant commissioning in line with the current industry practice.

Pressure vessel registration cards and similar data for other pressure containing equipment, which record the equipment manufacturing data and design conditions, should be prepared before plant commissioning and thereafter updated with inspection information and other data relevant to the continued safe working of the equipment.

3.3.2 Routine safety maintenance – frequencies

Safe operation of the plant relies, in particular, on the correct operation of reliable safety instrumentation, which demands regular inspection and testing. As a minimum, all instruments, relief valves, and major equipment controls, e.g. for compressors and their drives, should be checked and tested in the first year of plant operation. More frequent testing, usually monthly, will be required on certain defined safety systems to ensure that an acceptable standard of operation is achieved.

3.3.3 Conditions for safe entry and working under permit

Refer to section 1.2.9 for further information on safety control procedures.

3.3.3.1 Permits to work (clearance certificates)

3.3.3.1.1 Issuing and Receiving Authorities. Permits should be accepted by the man who is going to do the job. When several men will be doing the job, the permit should be accepted by the man in charge. In some companies it is the practice for a foreman to accept all permits. If this system is followed, the permit should be shown to the man or men who will do the job. A convenient way of doing this is to display the permit at the working area in a plastic envelope.

3.3.3.1.2 A typical permit. A good design of permit is illustrated in Figure 8:

(a) Part A is concerned with preparation.
   (i) Part A.1 is a checklist of hazards to be covered. Those not relevant are crossed out.
   (ii) Part A.2 states whether or not the equipment is isolated and, if so, by what means.
   (iii) If a fire permit (for hot work) or an entry certificate is required, this is recorded in A.4 and A.5. The fire permit or entry certificate should be attached.
   (iv) Part A.6 covers radioactive sources, A.7 electrical isolation, and A.8 records that preparation is complete. Note that section A is a checklist as well as a record.
(b) Part B lists the jobs to be done, the protective clothing required and any other precautions.
(c) Part C.1 is completed by the man preparing the equipment and Part C.2 by the man doing the job. Note that there is a space for four independent groups of men to work on the equipment. The rest of part C is completed when the permit is handed back.

All permits should cease to be valid after a stated period, say a maximum of 1 week (1 day when hot work entry is involved), and must then be handed back or renewed. The period of validity should be stated on the permit.

3.3.3.1.3 Isolation. Any piece of equipment on which work is to be done must be isolated from equipment containing hazardous materials. This is best done by the use of blinds (slip-plates) or by physical disconnection. Valves are liable to leak and may be opened in error, and therefore should not be used as the means of isolation.

Certain proprietary blinds which are used for product segregation may allow leakage past them if not positioned correctly in the line bore. These should not be used to isolate equipment for maintenance (except in circumstances in which valves would be accepted). Blinds used for isolation for maintenance should be the types which, if they leak, will leak into the atmosphere.

Where valves are used as the means of isolation, they should be locked shut, using a padlock and chain or similar device. This rule should apply to valves which have to be isolated so that blinds can be inserted (or disconnections made). When physical disconnection is used as a method of isolation, the lines leading to the rest of the plant should be blanked.

When a whole plant or section of a plant is shut down and freed from hazardous materials, it is not necessary for each item of equipment to be isolated individually. It is sufficient to make sure that the whole plant or section is isolated, as described above. Care is needed to make sure that process materials have been removed from 'dead-ends'.

Separate permits should be issued for:

(a) The initial isolation by fitting blinds or disconnection.
(b) The main job.
(c) Removing the blinds (or reconnecting the disconnected pipework).

Permit (b) should not be issued until job (a) is complete, permit (a) has been handed back, and the person issuing the permits has checked that (a) is complete and signed off. Permit (c) should not be issued until (b) is complete.

Additional precautions are necessary when equipment has to be prepared for hot work or entry (see section 3.3.3.2). Special care is needed when isolating relief valve or vent lines. Where possible, these should be disconnected, rather than blinded, in order to afford maximum protection to the equipment. If the arrangement of lines is such that disconnection is impossible, then the relief valve or vent line should be blinded last and deblinded first to ensure that the system cannot be overpressurized.

When electricity is connected to a piece of equipment, the supply is best isolated by one or more of the following means:

(a) Removal of fuses.
(b) Locking of isolators.
(c) Removal of breaker contacts, e.g. HV equipment.

An attempt should be made to start the equipment before work is begun, in order to check that the supply is isolated. The means of isolation must be positive and stated on the safety work permit.

Where any departure from established procedures is necessary, it should be carefully reviewed, and undertaken only when authorized in writing by a senior line manager.

3.3.3.1.4 Sweeping out and testing. Before equipment is opened up for repair, inspection, or modification, it must be freed, so far as is possible, from hazardous materials, and tested to confirm that they are absent.

The methods adopted for removing hazardous materials depend on their nature and the degree of hazard. Flammable or toxic gases at pressure should be blown off, if possible, to a flare or scrubbing system, or discharged from a vent stack at a safe height, and the equipment then swept out with inert gas or steam.

Sweeping out with inert gas can be carried out in two ways.

(a) By raising the pressure of the equipment to that of the inert gas supply and then blowing off. When pressurization is liable to lead to condensation of the material to be removed, special consideration should be given.
(b) By passing inert gas in at one side of the equipment and out at the far side.

Method (a) is best for equipment of irregular shape, where dead-ends might otherwise remain unswept. Method (b) is best in long, thin equipment, such as pipelines, provided that 'plug' flow can be guaranteed.

Hazardous liquids should be drained or blown out, preferably into a blow-down tank, another part of the plant, or into a closed drainage system. Water-soluble liquids are best removed with water but water must not be used with cryogens. Air can be used to remove non-flammable liquid. An inert gas may be safely used for blow-down in all systems.

Special precautions are necessary with materials such as benzene, which have a low threshold limit or high toxicity value. All pumps and other equipment on which maintenance is foreseen should have a closed connection to the drainage system.

Sweeping out normally takes place via purge points after the main valves have been closed and locked but before the blinds are inserted (or disconnections made).

After the cleaning operation has been carried out, it may be necessary to check that no hazardous materials are present. Special attention should be paid to dead-ends in the system, particularly if a whole plant, or section of a plant, is being prepared for maintenance and individual items of equipment are not being isolated individually. The tests will depend on the materials present. With flammable gases a test with a combustible gas detector is all that is usually necessary. With toxic gases a test for the specific gas may be necessary.

Special attention must be paid to testing when equipment is being prepared for hot work or entry.

The tests should be carried out immediately before work, at the precise location.

If water or steam is used to sweep out equipment, it will be necessary to make sure that no pockets of water remain. Water should not be used to flush stainless steel equipment unless the water is chloride-free, or stress corrosion cracking may occur.

3.3.3.2 Special permits

Special precautions are necessary for certain operations and these should be authorized on a special part of the permit form.

3.3.3.2.1 Entry to vessels and other confined spaces. Guidance is also given in Guidance Note GS5 from the Health and Safety Executive, 'Entry into confined spaces'. Isolation should always be by blinds or physical disconnection, and the blinds should be inserted or the disconnection made close to the vessel. There should be no valves or other restrictions between the blind or disconnection and the vessel, as process material may remain between the restriction and the blind. Where entry of personnel is required, valve isolations are not satisfactory even for quick jobs.

Any electrical equipment or machinery inside the vessel should be disconnected from the electrical supply, using the method outlined in section 3.3.3.1.3.

Special care is needed in sweeping out and testing. Vessels which have contained oils should be steamed for several hours and then filled with water (provided the vessel and supports can take the weight).

If entry is permitted without breathing apparatus, then the atmosphere at several points well inside the vessel, not just near the manhole, should be tested for oxygen content, flammable gases, and for other potentially hazardous materials, with particular reference to any known trace materials which might accumulate from the process. Entry without breathing apparatus should not be permitted if the oxygen content is below 20 per cent or if the concentration of any toxic gas or vapour is above the occupational exposure limit (OEL). Continuous monitoring of the oxygen content, using, for example, pocket-size oxygen alarms, is recommended.

Entry should not be permitted at all, even with breathing apparatus, if the oxygen content is above 22 per cent or if the concentration of flammable gas or vapour is more than 20 per cent of the lower flammable limit, as there may be pockets of higher concentration.

Entry permits should be authorized by a competent person of greater seniority than those who authorize normal permits. Permits should be valid for not more than 24 hours, and all tests should be repeated before renewal. The supervisor authorizing entry should personally inspect all connections before issuing the initial permit, and, at each renewal, make sure that they are blinded or disconnected and that any other special precautions necessary, such as forced ventilation, are in operation. He should also make sure that rescue facilities are available in case a man becomes injured or unwell while inside the vessel.

It may be necessary to issue a permit for 'entry for inspection' before a permit is issued for entry for work.

It is convenient to classify entry permits into one of three types:

(a) Type A permits are issued when the atmosphere is fit to breathe indefinitely. No special precautions are necessary other than those already described. However, another person must remain present at the entrance. Before issuing a type A permit, remember that disturbing scale, burning, welding, or painting may change the atmosphere inside a vessel. Special care is necessary if oxygen or compressed gas equipment is introduced into a vessel for welding purposes.
(b) Type B permits are issued when the atmosphere in the vessel contains some toxic or unpleasant gas or vapour, though not enough to present an immediate danger to life. Breathing apparatus must be worn and a lifeline should normally be worn. A person qualified in the use of breathing apparatus must be continuously on duty at the entrance to the vessel, and he must have additional breathing apparatus and resuscitation equipment available. He should also be able to summon assistance quickly, and rescue plans must be practised so that all concerned know what they have to do.

(c) Type C permits are issued when the atmosphere in the vessel is not respirable or contains so much toxic material that there is immediate danger to life. If possible, situations of this type should not be allowed to develop. On some plants, however, it may be necessary occasionally to enter a vessel which contains, for example, a catalyst which must not be exposed to air.

For a type C permit, in addition to the precautions for a type B, two persons trained in rescue should be on duty at the entrance to the vessel, and they should keep the person inside continuously in view. In addition, a safety harness/personnel hoist should be used to permit removal of the person from the vessel without entry. Where appropriate, means of rapid contact with the medical and rescue services should be available.

3.3.3.2.2 Hot work. Where the plant contains flammable materials, a special permit (usually known as a fire permit or hot-work permit) is necessary before any operations using sources of ignition are allowed. Further guidance is given in the Health and Safety Executive booklet HS (G) 5 'Hot work, welding, and cutting on plant containing flammable materials'. Such operations include welding, burning, use of non-classified electrical equipment, introduction of vehicles, and chipping of concrete. Before a hot-work permit is issued, two distinct hazards have to be considered: hazards due to the presence of flammable gas in the atmosphere, and hazards due to the presence of flammable materials in the equipment being worked on.

Permits for hot work should not be issued if abnormal plant conditions make a leak more likely than usual. The atmosphere should be tested with a combustible gas detector and no permit should be issued if more than 20 per cent of the lower flammable limit is present.

Whenever welding or burning is carried out, or cranes or other vehicles are operating in the plant area, portable combustible-gas-detector alarms should be placed nearby, and the persons concerned warned to shut off their welding equipment or engines if the alarm sounds. It is good practice to place several alarms around the place of work. As a general rule, particularly for welding and burning, a standby person should be in attendance with fire-extinguishers, hoses, or whatever equipment is considered appropriate.

On many plants, petrol- or diesel-engined vehicles are allowed on certain designated roads without special permission, but a hot work permit is required before they can leave these designated roads. There is no justification for treating diesel-engined vehicles as less dangerous than petrol-engined vehicles; diesel engines can ignite flammable gases unless they have been protected.

Special care is necessary before welding or burning is allowed near pools of water. Leaks of petrol or similar liquids hundreds of metres away may spread on top of the water and be set alight.

Care is also needed to ensure that flammable vapour cannot come out of the drains. This source of vapour is often overlooked, and drains should be checked for the presence of the flammable vapours before and during the maintenance operation.

Before welding or burning is allowed on equipment, it must be freed from flammable liquids, usually by steaming or by sweeping with inert gas, and then tested internally to make sure that any gas present amounts to less than 20 per cent of the lower flammable limit. Note that most combustible gas detectors will not detect flammable gas if no air is present; the sample may have to be mixed with air.

Whenever possible the equipment should be moved to a safe area before welding or burning is allowed. Welding should not be allowed on the *outside* of a vessel when people are working inside.

Service lines should always be tested for the presence of flammable vapour before welding or burning is allowed, because of the ease with which they can become contaminated with process material.

3.4 Storage systems

3.4.1 General

The physical properties of liquefied flammable gas fluids dictate the need to use materials of construction which are resistant to brittle fracture at low temperatures. Typical materials, for service at $-100°C$ and below, range from magnesium aluminium alloy, and 9 per cent nickel steel to stainless steels. Specific choice depends upon the application. Small volumes of LNG may be contained under pressure within a vacuum-insulated vessel, the vacuum being of the order of 0.01 mbar or less. Any leakage into the interspace would adversely affect the insulating properties. Hence potential leak sources are minimized where possible by the use of welded connections.

In contrast, a large refrigerated volume at near atmospheric pressure is contained within a double-walled, flat bottomed metal tank. The annular space between the two tanks is filled with special insulating material, such as expanded perlite.

The inner tank, in contact with the cryogen, is made of a suitable material, as described above, while the outer tank can be constructed of similar material, concrete or carbon steel plate, depending on the integrity requirements of the system. The outer tank contains the gas pressure and insulating materials.

Secondary containment is provided by a bund surrounding the storage tank. As a low earth bund offers little or no protection against a gas cloud spreading from a liquefied gas spill in the bund, some form of high level concrete bund, designed for cold shock from the release of the tank contents close to the storage tanks, is installed in modern storage designs. The bund can be a separate wall or an integral part of the tank design to form a double containment system. In addition, the whole system may be buried (excepting the roof) by provision of an earthen berm.

3.4.2 Ethylene storage

A riverside storage facility for import/export of liquid and gaseous ethylene by ship or pipeline is illustrated in Figure 11. The following safety features are incorporated:

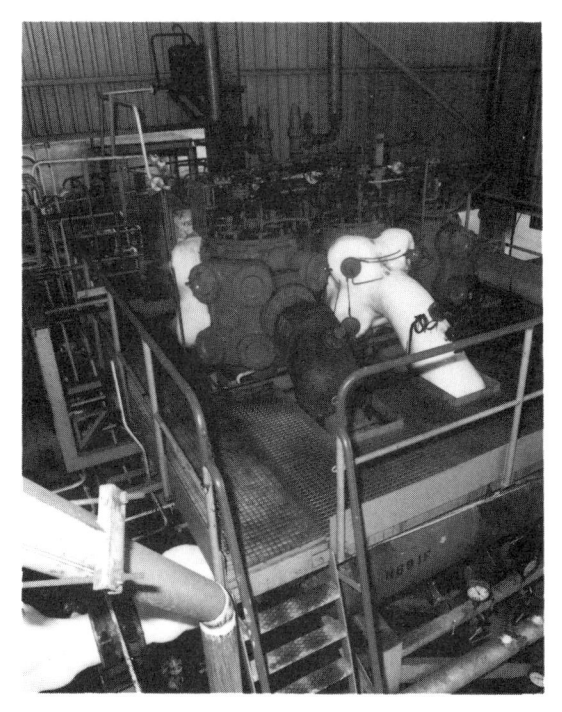

Figure 11 Ethylene storage tank at ICI's North Tees plant (*top*), and photograph inside the compressor house showing the ethylene compressors (note the build-up of ice) (*bottom*).

(a) A high level pre-stressed concrete bundwall to contain the total inventory of the storage tank. The narrow annulus between wall and tank restricts the surface area from which gas can be released in the event of tank failure, and the height above ground level minimizes the effect of flammable gas spread at ground level.
(b) The top bundwall has a steam curtain around the periphery in order to disperse a potential gas release to a safe concentration.
(c) Part way round the top of the bundwall is a water drench installation to provide protection against potential fires in adjacent storage facilities.
(d) The periphery of the storage installation is ringed by gas detectors to provide automatic warning in the plant control room of a gas release.
(e) A closed relief system is provided to collect all normal venting, purging and relief of flammable gas, which is then burnt safely in an elevated flare.
(f) An 'open' compressor house with open sides and roof-apex vent.
(g) The risk of liquid release from the storage tank is reduced by provision of weak seam roof sections. The design is such that in the event of total failure of the overpressure protective devices, a gas release from the roof will occur before the tank base holding-down straps fail. This minimizes the possibility of liquid release from the tank base.
(h) Ethylene content in the plant, outside the protected tank and bund, is minimized, firstly, by selection of an R22 refrigeration system and, secondly, by keeping the number of ethylene-containing plant items to a minimum.

3.4.3 LNG storage

Figure 10 shows an LNG storage tank (it may be noted that in this design the tank is surrounded by a high bund wall).

A large LNG facility to supply gas should incorporate the following features:

(a) Two approaches to bund design for containment of the total tank inventory:
  (i) A two spillage level arrangement, i.e., 10 and 110 per cent of the total tank contents.
  (ii) A high prestressed bund wall.
(b) High expansion foam monitors located at the 10 percent bund periphery to minimize the evaporation rate.
(c) Fixed water deluge facilities to cool the roof and wall of the tank exposed to heat from a possible fire in adjacent plant.
(d) Low temperature sensors situated within the bund for potential LNG spillage detection and activation of the foam system.
(e) Two approaches to liquid removal:

  (i) A single outlet line at the base of the tank protected by a fail-safe shut-off isolation valve, with remote activation in addition to the normal isolation valve.

(ii) In-tank pumps with 'over the top' outlet line, avoiding the need to breach the tank wall.

3.4.4 Large tank operations

3.4.4.1 Rollover

Rollover occurs when adjacent layers of LNG in a storage tank reach the same density and mix suddenly. The material in the lower layer is then under less head of pressure and consequently produces vapour rapidly. The total amount of extra vapour is proportional to the mass and supersaturation temperature of the lower layer.

In a 'normal' tank of LNG it is well established that there exists a very thin top layer whose temperature is slightly colder than the bulk of the LNG. Heat leak through the base and walls causes the adjacent LNG to be slightly warmer than the bulk, and this warmed LNG therefore rises to the surface, where it 'flashes' (creating boil-off). The LNG remaining after loss of boil-off is colder and therefore sinks through the bulk and mixes with it. This mechanism is aided by the density rise associated with the preferential evaporation of low molecular weight methane during boil-off. The tank is therefore in a state of dynamic equilibrium, the bulk of LNG being well mixed by these convection currents, while the surface layer is being constantly replenished.

If the tank pressure changes, the top layer responds virtually immediately, its temperature rising or falling accordingly. For example, if the boil-off compressor loading is increased, the pressure drops and the top layer temperature falls to a value which is in equilibrium with the lower tank pressure. The LNG rising to the surface now flashes off more than before (because the pressure is lower), and the colder LNG sinking into the bulk gradually results in a cooling of the bulk over a long period. Likewise, when the tank pressure is increased, there is less boil-off and the bulk LNG warms over a long period.

Over a long period of operation of an LNG tank the average boil-off is determined by the average heat leak into the tank. The top layer effect means that there can be variations in boil-off, the bulk of the LNG acting as a heat store or reserve of boil-off. The longer the period that boil-off is held below average by maintaining a high pressure, using a reduced compressor loading, the longer the period of high boil-off will tend to be when the pressure is reduced.

A prerequisite to rollover is the stratification of LNG in a storage tank. This may be caused by either of the following:

(a) The loading of fresh LNG into a tank containing a liquid residue (a 'heel'). This is more likely to cause stratification if a denser liquid is added underneath a less dense fluid, or a denser fluid is added on top of a lighter one.

Normally, liquid coming from the liquefaction plants is colder than the liquid in the tank, and hence is invariably denser (whatever the weathering, i.e. density vs time, characteristics). Denser liquids should always be top-filled to ensure good mixing.

The La Spezia rollover is the worst to have been recorded in the open literature. A 50,000m³ tank containing 43,000m³ of heavy Libyan LNG of specific gravity 0.54 suffered a rollover, with loss of 0.86 per cent of the tank contents. The average loss rate during venting was 0.47 per cent per hour.

(b) Autostratification. When LNG in a tank contains a higher proportion of nitrogen, greater than approximately 0.8 per cent, 'autostratification' is theoretically possible, as described by Chatterjee and Geist (ASME winter meeting, New York, 1976). As the LNG reaching the surface flashes, the boil-off vapour contains a high proportion of nitrogen. The LNG remaining, although colder, can be less dense because of the fall in molecular weight caused by loss of nitrogen. Hence there is no tendency for the top layer to sink into the bulk of the LNG. Under some circumstances therefore, the top layer, instead of being in dynamic equilibrium and being continually replenished, builds up as a stratified layer.

In this situation, the bulk of the LNG cannot reach the surface, owing to the layer above, and it gradually warms through heat leak into the tank. The top layer continues to boil-off, as it is warmed by heat transfer from the layer below, but at a much reduced rate.

Hence the pressure can be maintained with very low compressor loadings. This stratified condition cannot continue indefinitely. The top layer eventually loses so much methane that its molecular weight starts to rise again and its density increases (or at least decreases more slowly). The bulk of the LNG falls in density because its temperature is rising, and eventually the densities of top and bottom layers equalize. This is a metastable situation, and any disturbance, e.g. a change in compressor loading, will promote sudden mixing of the two layers. The bulk LNG, which was superheated, can now boil-off. This is a classic 'rollover', and results in a surge of boil-off over a period. The duration and rate of boil-off during a rollover are notoriously difficult to predict.

3.5   Firefighting

3.5.1   Fire situations

Identifications and risk of accidental fires on process plant will have been systematically studied at the project design stage. Some of the simpler aspects and principles are discussed below. Often other flammable liquids are present on a site as well as cryogens, and it is therefore important to appreciate that different fire situations may arise.

3.5.2   Fire initiation

It is the vapour phase of a liquid which takes part in the flame production. The readiness of liquids to form a vapour at ambient condition is therefore of prime importance. For example, an LNG liquid spill on the ground instantly creates a vapour at a high rate which then decays to a much slower rate, depending on heat leakage from soil/atmosphere. Following ignition, a short lived flash fire would occur, followed by steady burning from the pool. However, vapour evolution can be undesirably increased by breaking up a surface liquid pool, e.g. by fireman's jet hose, or if the liquid is discharged as a spray from a pressurized leak source.

In contrast, a flammable liquid such as diesel oil produces vapour much more slowly.

3.5.3 Fire extent

The pool diameter size gives guidance towards possible flame height. With most ignited spills, flame height will be limited to 1-2 times pool diameter. As the pool diameter increases, there is a tendency for the flame to break up into a number of smaller pool flames, as distinct from one main flame.

3.5.4 Fire duration

The duration of the fire is dictated by the regression rate and the depth of liquid.

3.5.4.1 Flammable leaks/delayed ignition

Serious consequences can result from small leaks, e.g. the escape from a failure of a 15mm line, if allowed to accumulate in confined environments. Should the accumulation be subsequently ignited, an overpressure or explosion could be created. Potential confined spaces include built-up process areas and spaces within buildings.

The importance of shutting off the fuel supply to the fire cannot be overemphasized. Equipment which experience shows is particularly liable to leak should be fitted with remotely operated emergency isolation valves.

3.5.5 Practical situations

3.5.5.1 Minor leaks

Leaks can occur during operation from pipe flanges, pump seals, and valve glands. In the design, consideration must be given to the possibility of a leak, ignition, and the consequences of flame impingement on associated plant.

3.5.5.2 Major leaks

Large quantities of hydrocarbon liquids need special consideration. For example, with a potential LNG tank fire it may not be possible to extinguish it, and hence the emphasis should be to protect surrounding plant. With storage tank farms, fixed water deluge systems may be provided on each tank, which will reduce the chance of a fire in one tank spreading to others.

Caution is needed with high pre-stressed concrete bunds, as an adverse external tank hydraulic overpressure could result if the deluge water was allowed to accumulate in the tank/bund interspace.

High-pressure storage introduces different requirements. The characteristics of a fire would be a torch-like flame. If the flame is vertical, the effects of thermal radiation will have to be considered. With a horizontal flame, spread to adjacent process equipment is a possibility.

Flames impinging on equipment may soften the metal and lead to a failure, with a subsequent boiling liquid expanding vapour explosion (BLEVE). In both situations priority must be given to isolating the fuel supply. Remotely operated isolation valves may be located on the inlet and outlet tank liquid lines to minimize spill duration.

A practice observed with some pressurized liquid storage systems is to provide a catchment pit situated away from the storage area to collect major spills. In this way it may be possible to transform a potentially serious fire situation into a controlled situation.

3.5.6 Firefighting agents and equipment

The two principal agents are water and foam.

Water is used for cooling adjacent plant in the vicinity of a fire and also for foam production. It is not a suitable firefighting agent for flammable liquids, but when used as fogging spray, it can be effective as a heat shield to assist access to the equipment near a fire.

Foam is the most effective means of suppressing and extinguishing flammable liquids when burning. For installations of 10,000 tonnes and above, a fixed foam distribution system should be installed.

3.5.6.1 Foams

Water in the foam provides a heat-resistant blanket of lower density than that of a flammable liquid, and is able to extinguish fire by:

(a) Cooling.
(b) Suppressing vaporization.
(c) Preventing the ingress of oxygen.

Foams are normally categorized in terms of their expansion ratio.

3.5.6.1.1 Low expansion foams. These are normally of the protein type and used for fighting fires where the process liquids are non-miscible in water, e.g. fuel oils, particularly those with a flashpoint below 100°C. Special 'alcohol-resistant' foam is required when dealing with water-miscible liquids, for which normal grades of foam would need extremely high application ratios.

3.5.6.1.2 Medium expansion foams. These are normally of the detergent type and have been found to be the best type of foam for fighting LNG fires. Compared to high expansion foam for fighting LNG fires, they provide very rapid blanket coverage over a large area. The foam compound has a good shelf life and requires less storage space than low expansion foams. Compared with low expansion foams, however, they have poor 'throw' i.e. cannot be projected any great distance from the generator. In addition, they are easily carried away by wind and are broken down by fire and by some chemicals.

3.5.6.1.3 High expansion foams. High expansion foams are particularly useful for extinguishing fires in large contained volumes, where the whole volume can be filled with the foam, i.e. by flooding buildings and other enclosures.

3.5.6.1.4 Aqueous film forming foam (AFFF). This foam is available in two forms – standard and alcohol-resistant. The experience with these foams is at present limited, but evidence currently available indicates that although they are more expensive than 'conventional' protein foams, the required application rates are lower; they are also less toxic and they do not deteriorate in storage.

3.5.6.2 Hand-held and mobile firefighting equipment

Advice and agreement of the local fire authority should be obtained in arriving at requirements to ensure compatibility with its equipment. Where equipment such as extinguishers, nozzles, etc., is kept out of doors, it is important that adequate weather protection is provided. The locations of all such fire points should be numbered and each location number shown on a layout drawing.

All equipment should be regularly inspected and maintained and a record kept of all such work. It should be noted that the occupier has a legal responsibility to do this.

3.5.6.2.1 Dry powder extinguishers. The preferred type of extinguisher for general use is the 10-12kg size of hand-held extinguisher.

All extinguishers should be permanently and distinctly marked to identify the type of powder. At specific locations on a site where 10-12kg extinguishers are not considered to be adequate, additional trolley-mounted 75kg units (or larger if the risk necessitates) should be provided. All continuously manned sites should be equipped with a minimum of two 75kg units. For installations which require a fire certificate, such as an LNG site, all dry powder extinguishers should be filled with a potassium powder (monnex or equivalent).

3.5.6.2.2 Monitors. For sites with low manning levels, portable water monitors should be provided where a firefighting main is installed. These monitors will enable cooling water to be applied on plant and equipment with minimal effort. A minimum of two such units should be available on each site.

3.5.6.2.3 Foam units. On sites where quantities of flammable liquids in excess of 2000 litres are stored, a minimum of one trolley-mounted foam unit of at least 100 litres foam-concentrate capacity should be provided.

3.5.6.2.4 Fire suits. A minimum of two fire protection suits should be provided on all permanently manned sites.

3.5.6.2.5 Firefighting equipment cabinets. Fire cabinets should be provided to store valve handles, hoses and nozzles, etc. Where it is necessary to provide a standpipe hydrant leg, this equipment should be housed in a suitable weather-protected cabinet.

3.6 Detection systems

3.6.1 Leak detection

Detection is normally used in order to give early warning of leaks, and is particularly important when manning is minimal. The effectiveness of a system can be determined only after careful consideration of the suitability of the sensing detector/s for the envisaged situation.

For pressurized fluid leaks, the movement of liquid/gas causes a change of energy. It may be possible to identify and detect this particular change of energy within the pipework or where the gas/liquid is leaking from the pipework. With LNG the 'cold' can be used as the basis of a detection system. In this instance, thermocouples or optic fibres are employed as sensing elements in the field.

The final physical change that occurs, in terms of time, is vaporization of the liquid to a gas and the formation of a flammable mixture with air. When a specific ratio of gas to air is present at the detector head (gas detection), the detection system becomes active.

3.6.1.1 Flammable-gas detection

Gas-detection systems have been developed for use in storage areas and plant areas where an undetected release of flammable gas or vapour could occur. A gas-detection system is basically used to give early warning of leakage, and is particularly useful in storage areas where operator cover is minimal, owing to the large area covered by the installations and the automatic action of the installed equipment.

3.6.1.1.1 General. In areas where more than one gas is to be detected the instruments should be calibrated for the least sensitive gas in that area.

Where the detectors depend on an effective catalyst, care should be taken to use the correct detector type to avoid poisoning of the catalyst. For example, in areas where silicone grease or oil is used, 'normal' detectors will require filter elements to be fitted.

As a general requirement, the operator should check that gases present, other than flammable gas being detected, are compatible with, and not likely to inhibit, the detector catalyst.

In practice, devices require fitting with draught cones and weather protection when mounted in the open.

As with most safety systems, the integrity of the gas-detection systems can only be maintained by regular testing of the equipment by trained personnel.

3.6.1.1.2 Gas-detection systems. There are basically three ways to use gas detection systems:

(a) The area is ringed on the periphery with gas detectors calibrated to read out percentages of the LFL (lower flammable limit) for the particular gas being measured. The detectors are spaced to cover the plant area, assuming that leaks have an angle of spread of 30 degrees.

Usually these detectors are connected to a control room annunciator panel, together with wind speed and direction indicator, such that an alarm is given and the operator can take the appropriate action, which can be to actuate emergency valves, call the fire brigade or activate steam curtains or other protective systems. Alternatively, the detectors can be used to start the protective systems automatically, such as steam curtains, to disperse the gas cloud.

Open-path infra-red sensors which can detect flammable gases at any point along their beams are being developed.

(b) Point source detectors may be used around equipment, such as pumps, separator pits, etc., which may have shown a high potential for developing gas releases, to give the operators, in a remote control room, early warning of danger. Another application is in analyser buildings, where the detector would be used to shut off automatically any power supplies feeding non-classified electrical equipment in the event of flammable gas being detected.

(c) Where maintenance work, especially hot work, is being carried out in a live plant area under a fire permit, portable gas detectors which give local audible and visible alarms may be used around the area of work to ensure that conditions remain safe for the duration of the maintenance activity, and that the work is stopped if the conditions change.

3.6.2 Fire detection

A similar approach can be made for the detection of fire. Important parameters are smoke particles, combustion products, and energy (convected and radiant heat).

Infra-red/ultraviolet detection in theory affords the fastest response, since the signal arrives at the detector instantaneously. Other types of detector can take the form of simple plastic tubes filled with air under pressure which burst at high temperature, release internal pressure and trigger an alarm system. In practice, a combination of detectors may be used after careful evaluation of the flame characteristics of the material in question.

Transmission of the alarm signal must be engineered for reliability and protected from foreseeable hazards, such as power failure, fire, explosion or vehicle impact.

Bibliography

'API Fire Safety Engineering Subcommittee LPG Task Force Report', September 1987. *Note*: This document formed the basis for API 2510/2510A.

API publication 2510A. 'Fire Protection Considerations for the Design and Operation of Liquefied Petroleum Gas (LPG) Storage Facilities' (1989).

API Standard 2510. 'Design and Construction of Liquefied Petroleum Gas (LPG) Installations' (1989).

Chatterjee, N. and Geist, J. ASME Paper 76-WA/PID-6 (1976).

HS/G 34. 'The storage of LPG at fixed installations' (1987).
'Liquefied Flammable Gases Storage and Handling. Engineering Codes and Regulations', Group D, vol. 1. 6, Imperial Chemical Industries (1970).
Liquefied Petroleum Gas, vol. 1, 'Large Bulk Pressure Storage and Refrigerated LPG. Model Code of Safe Practice, Part 9', Institute of Petroleum (1987).
NFPA 58. 'Standard for the Storage and Handling of Liquefied Petroleum Gases' (1989).
NFPA 59. 'LP-Gases at Utility Gas Plants' (1989).

4 Hydrogen

This part supplements the general safety requirements in Chapter 1 to cover the special precautions to be observed in the operation and maintenance of plant and equipment used in the handling of liquid and gaseous hydrogen at low temperatures. The greatest hazards are those of fire or explosion. Other hazards, resulting from high pressures, low temperatures, and atmospheres other than normal, will be referred to as secondary hazards.

4.1 Specific hazards

4.1.1 The fire and explosion hazard

4.1.1.1 General

Hydrogen is a high energy fuel which, when mixed with air or other oxidizers, releases large amounts of energy in the form of heat or explosive force when ignited.

The gas is very light and travels rapidly upwards at ambient temperature. Release under eaves outside a building can therefore present a problem, as the gas may find its way into the building. Within buildings, high-level ventilation should always be provided.

Special care is necessary in any part of a process necessitating possible operation below atmospheric pressure, as this can quickly lead to the formation of an explosive mixture due to ingress of air.

4.1.1.2 Ignition and burning characteristics

4.1.1.2.1 Spontaneous ignition temperature (SIT). Hydrogen has a spontaneous ignition temperature of about 450°C (723K), but this value depends on measurement conditions.

4.1.1.2.2 Laminar burning velocity and significance of confinement. The laminar burning velocity is relevant to the violence of explosions and to the tendency of a flame to 'run up' into a detonation. The burning velocity of hydrogen is markedly greater than that of other common fuels, indicating that the hazard of explosions and detonations must be carefully considered. A practical consideration that affects the probability of 'run up' is the degree of confinement of the flammable mixture; the greater the degree of confinement, the greater the probability of explosion. It has been shown that even with partial confinement, burning to detonation of hydrogen/air mixture can be expected.

4.1.1.2.3 Minimum ignition energy. Spark ignition energies may be used as a measure of sensitivity to ignition by local sources. Hydrogen differs from many other flammable gases in that it requires significantly less energy for ignition, e.g. in air, the minimum ignition energy for methane is 0.28 millijoules, as opposed to 0.02 millijoules for hydrogen. This is borne out by the fact that venting or leaking hydrogen can ignite spontaneously. The presence of dust in a vent stack can cause ignition of hydrogen by discharge of static electricity.

4.1.1.2.4 Flammable range. Appreciation of the wide range of concentrations in air or oxygen in which hydrogen is flammable is very important. The flammable range at 1 bar of hydrogen in air is 4–74 per cent by volume; in oxygen it is 4–94 per cent. Compared to other flammable gases, this is very wide. Note that methane has a range of 5–14 per cent in air; for propane, the range is 2.1–9.5 per cent. At pressures and temperatures above ambient or with inert gases other than nitrogen, e.g. argon, the range can be widened.

4.1.1.2.5 Limiting oxygen index. The limiting oxygen index is the minimum concentration of oxygen to support flame propagation when a stoichiometric fuel–air mixture is diluted with nitrogen. In other words, it is the least amount of oxygen required to support combustion. In general, gaseous hydrocarbons require 10-12 per cent oxygen by volume in the original atmosphere to support combustion, but hydrogen requires only 5 per cent oxygen.

This value, of course, sets a limit on the quantity of oxygen that can be tolerated when purging or inerting equipment. (Note: air contains 21 per cent oxygen by volume.)

4.1.1.2.6 Quenching distance. This is the minimum gap between two parallel surfaces which will just permit a flame to pass without cooling it to extinction. The smaller the quenching distance, the greater resistance the flame has to cooling, and the more difficult it will be to extinguish. The quenching distance for hydrogen is about one-third that for hydrocarbons such as propane. Thus, flameproof equipment for hydrogen has to be designed to a much higher standard than for other flammable gases.

4.1.1.2.7 Flame characteristics. Hydrogen burns at higher temperatures, but generally gives off much less radiant heat than propane or other hydrocarbons. The flames are colourless or nearly colourless. Both these characteristics make it more difficult to detect a hydrogen fire. Detection of burning hydrogen is easier at night or in subdued lighting. Ultraviolet sensors, linked to alarm systems, can be used to detect and give warning of the presence of hydrogen flames.

4.1.1.3 Oxidants and their avoidance

The oxidizing agents which may exist in a hydrogen plant are air, cold-box atmospheres containing air diluted with nitrogen, or oxygen-enriched air. It should be remembered that air will initially be present in cold boxes, process vessels, and pipework. When air condenses on cold equipment, the liquid produced is substantially enriched in oxygen and is therefore a more powerful oxidant than air.

4.1.1.3.1 Purging. The limiting oxygen index is relevant only to the problem of flammability. Considerations such as solubility, liquefaction, or freezing at the process temperature involved, may require limiting the oxygen to much lower concentrations than those dicated by flammability alone.

Before the introduction of hydrogen into processing equipment, air should be eliminated by nitrogen purge to give an exit oxygen concentration of less than 20 per cent of the minimum oxygen index.

4.1.1.3.2 Air ingress. Hydrogen plants should never be operated below atmospheric pressure, as this could lead to the ingress of air and the formation of flammable mixtures. Suitable safeguards should be provided to prevent operation below atmospheric pressure.

4.1.1.3.3 Oxidant removal. Hydrogen feed gas that may contain oxygen should have the oxygen removed catalytically at ambient temperatures. In addition, low-temperature adsorbers should be provided to remove trace quantities of oxygen or other oxidizers where concentrations warrant it. At liquid hydrogen temperatures, oxygen is soluble in liquid hydrogen to the extent of only a fraction of a part per million, so small concentrations of oxygen can cause accumulations of solid oxygen with a potential explosion hazard.

Other potential oxidants such as nitrogen oxides should be removed by adsorption, catalyst, or other means.

4.1.1.3.4 Monitoring. Analysis by low-temperature adsorption and chromatography or other means should be provided to give maximum assurance of freedom from potential hazards due to contaminants.

4.1.1.4 Condensation of air and other gases

The low temperature of liquid hydrogen can solidify any gas except helium. The solidified gases could plug restricted areas, such as valves or small openings, and cause equipment failure. If air or oxygen is allowed to condense and solidify in liquid hydrogen, a potential explosion hazard can result. Because air (oxygen) will condense into it, liquid hydrogen should generally be handled in closed systems provided with safety relief devices which exclude air. An exception may be made where, with proper precautions, as outlined below, quantities of a few litres of liquid hydrogen are handled in laboratory or test operations in open Dewar vessels properly stoppered and vented.

The venting systems on liquid hydrogen containers should be examined periodically to make sure that they do not become plugged with moisture frozen from the air.

4.1.1.5 Electrostatic effects

Electrostatic charges can build up as result of friction in pipes or by the break-up of liquid hydrogen to droplet size. This can cause ignition where gas is venting to atmosphere.

4.1.1.5.1 Safeguard against electrostatic build-up. Routine inspection to ensure the integrity of electrical earthing and bonding systems on plant equipment is essential. When maintenance work has been carried out, the electrical bonding should be checked before recommissioning.

4.1.1.5.2 Road tanker operation. Before transfer begins, tankers carrying liquid hydrogen must be equipped with an earth cable for connection to the earth terminal of the receiving equipment, and remain connected until transfer is completed.

4.1.1.5.3 Electrostatic ignition risks. When gaseous hydrogen and air are mixed within the limits of flammability, the threshold ignition energy can be as low as 0.02 millijoules.

The maximum electrostatic energy which can be acquired by a man standing on an earthed plate in insulating footwear is about 15 millijoules; thus it is desirable to dissipate this body charge. This can conveniently be done by wearing anti-static, non-sparking footwear having a leakage path of not more than 5 megohms to earth.

4.1.1.6 Electrical equipment

All electrical equipment, such as switches, motors, instruments and portable testing devices, used and installed within the boundary of the hydrogen installation shall be suitable for the requirements of the area classification. The classification of areas will be in accordance with the definitions in BS 5345, parts 1 and 2.

4.1.1.7 Firefighting

Hydrogen fires burn with an almost invisible flame. Hence the particular need to fight from a safe distance and from the upwind side.

The most effective way of dealing with a hydrogen fire is to shut off the hydrogen supply. Depending upon the circumstances, it is not usually advisable to extinguish a hydrogen flame in confined areas if the hydrogen supply cannot be shut off. The continued escape of unburned hydrogen can create an explosive mixture which may be ignited by other burning material or hot surfaces. It is better to allow hydrogen to burn in localized areas and keep adjacent objects cool with water rather than risk the possibility of an explosion. Proceed as follows:

(a) Use large quantities of water, preferably in the form of a spray, to cool adjacent equipment and to cool any burning material below its ignition point. Do not apply directly to a pool of burning liquid hydrogen, as water will evaporate additional gas.
(b) Liquid hydrogen when exposed to the atmosphere will produce a cloud of ice/fog from the air. The flammable mixture will probably extend beyond this vapour cloud, and personnel should therefore keep well outside the area of visible cloud.
(c) If electrical equipment is affected by the fire, disconnect the supply or use carbon dioxide or dry chemical extinguishers above and not directly on to the fire.

4.1.2 Secondary hazards

These take the following forms:

(a) Mechanical hazards arising from the very low temperature at which hydrogen is processed.
(b) Health hazards applicable to hydrogen.
(c) Health hazards due to other gases associated with hydrogen processing.

4.1.2.1 Mechanical hazards

These hazards may take the following forms:

(a) Pressure rupture – due to trapping cold liquid or vapour.
(b) Brittle fracture – due to metals being used below their embrittlement temperatures.
(c) Freezing leakage. If water is present before cooling down, plant and equipment fractures may occur, owing to the expansion of ice formed. Plant dryness is therefore vital.
(d) Contraction leakage. Sudden cooling down can lead to contraction and leakage from bolted flanges, which may have to be specially designed to overcome this problem.

These hazards have been dealt with in more detail in Chapter 1 but it should be remembered that, owing to the very low temperatures involved in liquid hydrogen plants, the effects of the various hazards are liable to be more pronounced than they would be in the case of other gases, except helium.

4.1.2.2 Health hazards – hydrogen

The hazards to health of gaseous hydrogen and liquid hydrogen stem from their low temperatures and the exclusion of oxygen arising from vaporization in enclosed spaces.

Serious tissue destruction similar to that caused by burns can occur when liquid hydrogen comes into direct contact with body surfaces, owing to the low temperature of the liquid. Similar effects can arise from body contact with uninsulated equipment containing cold fluids.

Hydrogen is non-toxic, but oxygen exclusion will result in asphyxiation, and it should be noted that this can often occur so rapidly that the victim may be unable to escape even if he realizes that something is wrong.

4.1.2.3 Health hazards – other gases

In addition to hydrogen, aromatic hydrocarbons, hydrogen sulphide, carbon monoxide, and nickel carbonyl are often present in the purification section of a hydrogen plant.

The aromatic hydrocarbons are poisonous as well as asphyxiating. Acute poisoning resulting from the inhalation of high concentrations of vapour (above 3000 ppm benzene by volume in air), occurs over a few minutes.

4.1.2.3.1 Hydrogen sulphide. A highly toxic gas. At low concentrations hydrogen sulphide acts as an irritant to the eyes and nose. In these concentrations it is easily recognizable by a characteristic rotten eggs smell, but after exposure the sense of smell is dulled, making it an unreliable means of warning of exposure. Exposure to moderate concentrations causes headache, dizziness, nausea, and vomiting, in that order, followed by loss of consciousness, respiratory failure, and death.

4.1.2.3.2 Carbon monoxide. A particularly highly toxic gas for which the occupational exposure limit (OEL) is 50 ppm. The gas is colourless, odourless and tasteless, and can form flammable, and possibly, explosive, mixtures with air. The flammable range in air is 12.5–70 per cent by volume.

Carbon monoxide acts as an asphyxiant. Its action is to combine with haemoglobin in the blood to form a stable compound which does not react with oxygen and thus prevents oxygen transport by the bloodstream.

Personnel entering an area where there may be concentrations of carbon monoxide must wear adequate protective devices, such as a positive-pressure respirator. Where concentrations of carbon monoxide are likely to exist, constant or periodic monitoring of the atmosphere should be carried out.

In the event of a person showing symptoms of carbon monoxide poisoning, the following action is recommended:

(a) Ensure that personal risk is minimized, immediately remove the victim to a safe area.
(b) Ensure that there is no obstruction of the airway. If breathing is weak, apply artificial respiration with simultaneous administration of oxygen using an oxygen resuscitator.
(c) Summon medical asistance.
(d) Keep the victim warm and relaxed.

4.1.2.3.3 Nickel carbonyl. This is also a highly toxic compound, even in low concentrations. The maximum allowable concentration in air is 0.002 ppm. It is also a flammable material and a dangerous fire and explosion hazard. Before maintenance, a check should be made with detector tubes to ensure that it is not present.

4.2 Safety in maintenance and operation of hydrogen plants

4.2.1 Definition of hydrogen plant

The term 'hydrogen plant' is used to define any plant designed primarily for the low-temperature separation and/or liquefaction of hydrogen.

Such plants have both hot and cold sections, and it is not intended in this guide to define general safety requirements associated with those sections operating at ambient temperature or above. However, where contamination of these sections of the plant could lead to contamination of the cold sections, special precautions, as described below, must be observed.

The low-temperature construction in which the cryogenic process equipment is contained is known as the 'cold box'. It requires the same general construction considerations as other cryogenic process plants, except that a flammable product is being dealt with, and temperatures considerably lower

Hydrogen 83

than the liquefaction points of oxygen and nitrogen are encountered. These factors must be taken into account.

4.2.1.1 The cold box

The 'cold box' consists of the following basic components:

4.2.1.1.1 The cold box structure. This is generally a large box or cylinder. It is constructed from mild steel plate and structural members and it has the following functions:

(a) To support the process equipment.
(b) To contain the loose-fill insulation.
(c) To maintain the protective inert atmosphere.

4.2.1.1.2 The process equipment. This can include heat exchangers, vessels, columns, pumps, expanders, valves, piping, and instruments.

4.2.1.1.3 The thermal insulation. Typical examples are expanded perlite powder, mineral wool, or vacuum, aluminium foil, multi-layer super-insulants, or a combination of expanded perlite and vacuum.

4.2.1.2 Plant external to the cold box

This comprises the following:

(a) Process equipment: adsorbers, heat exchangers.
(b) Machinery: compressors, pumps, expanders.
(c) Relief valves, relief header, stacks, flares.
(d) Blow-down system pipework and vessels.
(e) Purging and venting systems.
(f) Pipework and valves.
(g) Control instruments and valves.
(h) Gas or liquid trailer fill points.

4.2.2 Safety in maintenance of hydrogen separation or liquefaction plant

4.2.2.1 Cleanliness

It is important to prevent moisture, oil, or grease entering the cold sections of the plant, where they could cause sticking of valves or blockages of pipework and equipment.

Various sections of the cold box operate at temperatures below the condensation temperature of the air. If the purge should fail and air enter the cold box, condensation of oxygen-enriched air can occur on the outside of cold equipment. It is therefore important that equipment be kept clean, both internally and externally.

For the same reasons a high standard of cleanliness must be upheld for equipment associated with the cold box during maintenance operations.

4.2.2.1.1 Special maintenance area. An area for final cleaning and assembly of hydrogen plant components must be clearly differentiated from other maintenance areas, and kept in a clean and oil-free condition. Combustible materials must be kept to a practicable minimum in this area.

Dismantling and initial cleaning should be carried out before components are taken to the clean area for repair, final cleaning, and reassembly. However, the clean maintenance area need not be clean to surgical standards, e.g. does not require air-conditioning, entry locks, etc.

4.2.2.1.2 Tools. Tools for erecting and maintaining hydrogen plant should be kept thoroughly clean. Recent practice has tended to favour the use of steel tools, as opposed to non-sparking tools, coupled with more stringent safety precautions.

4.2.2.1.3 Protective clothing. Protective clothing such as overalls, footwear, gloves, etc., solely for use when handling hydrogen plant, should be kept free from oil or grease. Greasy overalls, which have been used during general maintenance work, must be thoroughly cleaned before being worn while maintaining hydrogen plant.

In general, anti-static overalls, footwear, and gloves are recommended.

4.2.2.1.4 Personal cleanliness. Personnel should cleanse hands thoroughly and should change into approved protective clothing before transferring from any general maintenance involving dirt, oil, or grease, etc., to plant maintenance. Barrier protection creams should not be applied to the hands before handling hydrogen plant components which have already been cleaned for service.

4.2.2.2 Ventilation of buildings

Ventilation should be provided in buildings where flammable gases or liquids are handled, to prevent the hazardous accumulation of flammable vapours.

For gases lighter than air, roof ventilation is essential.

4.2.2.3 Safety control procedures

The information on safety control procedures given in sections 1.2.9 and 3.3.3 are applicable to hydrogen plant and equipment.

The physical properties of hydrogen pose specific problems which must be recognized when preparing equipment for repair or maintenance. Particular attention must be paid to the following precautionary measures:

4.2.2.3.1 Purging of vessels and confined spaces. Hydrogen is highly flammable over a wide range of concentrations in air. Before personnel enter a cold box, process vessel or other confined space which has been in service with hydrogen or hydrogen-enriched gas, it is essential that the systems are first purged with nitrogen (or other inert gas) to remove all flammable gas and, subsequently, with air. Before entry the atmosphere should be analysed to ensure that it is not deficient in oxygen.

4.2.2.3.2 Equipment isolation. Hydrogen is the lightest known gas and the most searching. Consequently, valves and mechanical joints in piping systems in service with hydrogen are more prone to leakage than those in service with heavier gases such as air, nitrogen or oxygen. Before starting any maintenance work, particularly if welding or flame-cutting operations are to be carried out, it is essential to isolate the equipment from other parts of the plant which may still be in service. For the reasons given above, valves cannot be regarded as a reliable means of isolation in hydrogen systems. Equipment isolation must therefore be effected in a positive manner by removing spool pieces or inserting blind flanges.

4.2.2.3.3 Sources of ignition. Hydrogen ignites very easily because it requires a much lower energy for ignition than many other flammable gases. For this reason it is important to ensure that sources of ignition such as open flames from cutting and welding torches or sparks from electric power tools or other sources are prohibited from the hazardous zones defined in section 4.2.2.3.4 below, unless the work area is declared safe by the issue of a work permit.

Before hot work is carried out on or adjacent to the external surfaces of vessels or piping which contain hydrogen and other flammable gases, the equipment must be purged with nitrogen as a minimum requirement and declared as being in a safe condition.

If hot work using open flames or sparks is to be carried out within 50ft (15m) of a hydrogen plant, the surrounding atmosphere must be checked with a flammable gas detector both before and during the work activity to ensure that hydrogen is not present. Hot work must be prohibited where hydrogen and other flammable gases are being vented.

Before an area is declared safe for hot work, it is important to ensure that sparks or molten metal cannot be projected or dropped into a potentially hazardous area. This applies particularly to work carried out at high elevations.

4.2.2.3.4 Work permit areas. Because hydrogen is highly flammable, a work permit must be issued by a responsible person before entry inside the following areas or before items of equipment are allowed for maintenance that employs either hot or cold work:

(a) Cold boxes, vessels, tanks or gasholders.
(b) Pits, trenches or ducts within 100ft (30m) of the hydrogen plant or product storage area.
(c) Any other enclosed or confined space where a possibility of enrichment by hydrogen or other flammable gases exists.
(d) A zone within 50ft (15m) of a plant processing or producing hydrogen.

4.2.2.4 Sources of ignition during maintenance: gas or electric welding, cutting or heating equipment

Such equipment must not be used for plant maintenance in any danger area as defined in section 4.2.2.3.2 until the atmosphere has been certified as safe by the factory manager by issue of a 'permit to work'.

When a flame operation is to be carried out on the exterior of vessels or piping which may contain hydrogen or hydrocarbons, these are to be purged with nitrogen and a sample of the gas contents analysed; and the equipment should be specified in the 'permit to work' as being in safe condition on which to work.

Before operating cutting and welding blowpipes or torches within 50ft (15m) of hydrogen separation or product-handling plant, extreme care should be taken by using a gas detector to ensure that hydrogen concentrations are not present and do not become present while work is being carried out. Flame operations are prohibited where gaseous or liquid hydrogen or hydrocarbons are being vented.

Even in cases where a particular area has been certified as safe for welding or cutting operations, precautions should be taken to ensure that no sparks, oxide, or molten metal, etc., can be projected or dropped into an unsafe area. This particularly applies when operations are being carried out above ground level.

4.2.3 Safety in operation of hydrogen separation or liquefaction plant

The following guidance for plant supervisors, operators or attendants applies to plant as defined in section 4.2.1.

4.2.3.1 Unsafe materials near hydrogen plant

4.2.3.1.1 'Unsafe materials danger zone' hydrogen plant. All space within 50ft (15m) of hydrogen separation or product-handling plant, including the air space above, should be regarded as 'unsafe materials danger zone'.

4.2.3.1.2 Unsafe materials in danger zone. Unless absolutely unavoidable, combustible material should not be installed, stored, or temporarily placed in unsafe material zones as defined in section 4.2.3.1.1 above. Materials such as wood, cleaning cloths, rags, and containers of oil, petrol, and paint are portable and can easily be excluded from these zones.

It is recognized that some combustible materials cannot be excluded completely from these zones, e.g. paint for protection of the plant, oil in machines, personal clothing, liquid transport vehicles. Therefore this section gives guidance only to prevent avoidable and unnecessary hazards.

4.2.3.2 Prohibition of smoking and use of naked lights

4.2.3.2.1 Flame and no smoking danger zone near hydrogen plant. A danger zone has been defined in section 4.2.3.1.1 above. Smoking and use of naked lights are prohibited within this zone while the plant is in operation or contains any hydrogen or hydrocarbon vapours. However, an extended 'no smoking' danger zone may be defined, as deemed necessary, by the factory management to embrace any additional area or the entire plant site, for the purpose of preventing danger from fire, heat, or smoking.

4.2.3.2.2 Danger from fire, heat, or smoking. The dangers from gas or electric welding or heating equipment, smoking and naked lights during maintenance work have already been defined in section 4.2.2.4. These must be known to plant operators and attendants, and if maintenance is being carried out while the plant is in operation, the operator should be aware of the situation and should check that rules are being observed by maintenance staff and that 'permits to work' are issued when necessary.

4.2.3.3 Action in regard to leaks

4.2.3.3.1 Detection of leaks. Analysis equipment will generally provide the first means of confirming that a leak is present and of showing the nature of the gas.

Where the source of leakage is not obvious, a liquid detergent diluted with water can be used for locating those on external joints or piping. New frost patches that appear on the outside of cold boxes or other low temperature equipment usually signify that there is an internal leakage of cold gas or liquid.

4.2.3.3.2 Notification of leaks. When there is evidence of leakage in a hydrogen plant or from external lines, a competent person must be notified immediately, and he will decide whether the plant should be shut down or not.

4.2.3.3.3 Areas affected by leaks. Areas affected by leaks must be regarded as 'danger areas', as defined in section 4.2.3.1.1, and the necessary precautions must be taken.

4.2.3.4 Safety devices

Reference should be made to section 1.2.1 of this manual.

Safety valves and rupture discs are for emergencies, and with proper equipment operation they will seldom be called upon to discharge gas.

(a) Safety valves, rupture discs, and blow-down vents discharging hydrogen or hydrogen-rich gases should discharge to a safe area.
(b) It is suggested that where previous experience is not available, then relief valves should be checked at 12-monthly intervals. Based on satisfactory operating experience, extensions to 24 months could be permitted if agreed with the insuring authorities.
(c) Relief valves and rupture discs should be regularly inspected for leakage and frosting.
   Note. Relief valves feathering below their set pressure, or leaking relief valves, will lead to excessive ice formation on the valve body, which can prevent the proper operation of the valve and plugging of the discharge line, rendering the valve inoperable in an emergency.

4.2.3.5 Cold box atmospheres

(a) Cold box atmospheres are to be analysed frequently.
(b) Any changes in the analysis must be reported to a competent person immediately.
(c) If there are any leaks of major proportions, as evidenced by quick-forming large frost spots, clouds of vapour, or unaccounted internal loss of liquid, advise the competent person immediately, and he will decide whether an immediate shutdown is necessary.
(d) The design cold box purge flowrate and pressure should be maintained at all times.
(e) Pressures above the operating limits will lead to discharge of insulation through the blow-off valves. Pressures below the operating limits can lead to the ingress of moisture and air.

(f) Equipment operating below the liquefaction temperature of nitrogen must be contained in a cold box which maintains a hydrogen atmosphere or a vacuum. If a hydrogen atmosphere is used, air should be removed by nitrogen purge, followed by the introduction of hydrogen. The hydrogen atmosphere must be maintained at a pressure above atmospheric at all times to prevent the ingress of air.

Before maintenance, a hydrogen atmosphere should first be removed by purging with nitrogen before introducing air.

4.2.3.6 Purging of equipment before start-up

After a plant shutdown, any equipment that has been opened will be filled with air, and it is imperative that this air is removed from the plant before hydrogen-enriched gases are introduced.

Air must first be purged out of the equipment, using dry nitrogen or helium. The nitrogen/helium should then be removed by purging the plant with feed gas.

Helium, although more expensive than nitrogen, has the advantage that it does not condense at liquid hydrogen temperature.

4.2.3.7 The disposal of liquefied gases

(a) Hydrogen separation and production plants contain quantities of liquefied gases held up in process vessels. When a plant is shut down, it may be necessary to dispose of these liquefied gases.
(b) Plant manufacturers generally supply liquid disposal units which form part of a blow-down system; these disposal units enable quantities of liquefied gases to be disposed of safely.
(c) The liquid blow-down system consists of drain lines, drain valves, vaporizers, vent stack, purge system and controllers.

The system should be operated in accordance with the manufacturer's instructions.

4.2.3.8 The venting of liquefied gases trapped in pipelines and equipment

High pressure can be generated when liquid hydrogen or light hydrocarbons are trapped in a closed system (see section 1.2.1). Generally relief valves or bursting discs should be provided to prevent this. However, it is good practice to blow-down such systems manually.

4.2.3.9 Emergency procedures

Emergency procedures should be prepared in consultation with local authorities to cover fire or any other hazardous event which may occur. Periodic emergency drills should be carried out.

The following guidelines may be used for formulating emergency procedures:

(a) Raise the alarm.
(b) Actuate fire water systems to keep equipment cool.
(c) Summon help and emergency services.
(d) Isolate the source of hydrogen, if appropriate and safe to do so.

(e) Evacuate all persons from the danger area and seal it off.
(f) Alert the public to possible dangers from vapour clouds, and evacuate when necessary.
(g) Notify the gas or plant supplier immediately.

4.3. Safety in maintenance and operation of liquid hydrogen storage, transport and handling equipment

4.3.1 Definition of liquid hydrogen storage, transport and handling equipment

This equipment includes all component parts, assemblies, sub-assemblies, piping, and instruments in contact with the gaseous or liquid hydrogen, including:

(a) Storage and transport tank inner vessels, piping, safety valves, fittings, and instruments.
(b) Liquid pumps, suction piping, high pressure piping, safety valves, instruments, excluding the oil-lubricated motion work.
(c) Evaporator inner vessels, evaporator coils, connecting piping, valves safety valves, fittings, and instruments.
(d) Piping and valves for liquid storage manifold system.

4.3.2 General

It is essential that personnel concerned with maintenance and operation of liquid hydrogen storage, transport, and handling equipment should be familiar with Chapter 1 of this guide. The points laid down therein are applicable and the following text covers only such additional features as require emphasis.

4.3.3 Operation of liquid hydrogen storage and handling equipment

Liquid hydrogen storage and handling equipment should be designed and installed in accordance with established codes and safe practices which incorporate all necessary features for their safe operation. A bulk storage tank for liquid hydrogen is illustrated in Figure 12.

4.3.3.1 Storage and loading area

Operators should be properly trained in the operation of equipment and in dealing with emergencies, e.g. spillage, fire.

4.3.3.2 Transfer lines

Liquid hydrogen is normally transferred through vacuum-jacketed piping systems to minimize heat leak and thereby conserve liquid product. To maintain the integrity of the piping system and limit the exposure to fire damage, combustible materials should not be stored or deposited under transfer lines which carry liquid hydrogen.

4.3.3.3 *Identification*

The line system and the fluid being handled should be clearly identified at all times.

4.3.3.4 *Access to loading area*

Clear access for vehicles should be provided. Hard standing for plant and for filling vehicles should comprise impermeable non-combustible material.

4.3.3.5 *Ventilation*

In covered areas adequate ventilation must be maintained. Where possible, installations should be located outside.

4.3.3.6 *No smoking area*

Areas for liquid hydrogen storage must be adequately guarded and posted to prevent access by unauthorized personnel and must be defined as a 'no smoking, no naked lights' areas.

4.3.3.7 *Tow-away accidents*

There should be procedures for preventing road tankers or trailers driving away from fill points with the transfer hoses still connected. Such incidents can result in equipment damage and an uncontrolled release of liquid or gaseous hydrogen, giving rise to a serious potential fire hazard.

4.3.4 Materials of construction for liquid hydrogen storage and handling equipment

Storage, transport and transfer equipment will normally be fabricated from austenitic stainless steel, aluminium, copper, monel, or other material having satisfactory low temperature properties.

Most ordinary steels are very brittle at low temperature and are not satisfactory. It is therefore essential that any modifications are properly engineered and approved by the appropriate design authority.

4.3.5 Purging

The system of purging hydrogen storage and handling equipment is identical to that described in section 4.2.3.6.

4.3.6 Protective clothing

During transfer operations, personnel should wear gloves, and goggles, or preferably full faceshields, to prevent liquid hydrogen contact with the skin. Cold burns may result by contact of liquid hydrogen with the skin after only a few seconds. If liquid hydrogen contacts the skin, irrigate the affected area with cold water and obtain prompt medical attention. The wearing of flame-resistant overalls can provide additional protection in the event of fire.

4.3.7 Loading and transfer areas

The following precautions should be observed during loading and transfer operations:

(a) Eliminate sources of ignition, e.g. smoking, open flames, and sparking devices of any nature.
(b) A gas-detector reading should be taken to confirm a safe atmosphere exists.
(c) Access should be restricted to authorized traffic and personnel.
(d) Established transfer procedures should be strictly observed.

4.3.8 Maintenance of liquid hydrogen storage, transport and handling equipment

A preventative maintenance programme should be established on the basis of operating experience and in consultation with the equipment manufacturers.

Cleaning of tanks is generally required before first commissioning only, because the product is pure, dry, and non-corrosive. Cleaning may not subsequently be needed, unless there is known to have been severe contamination due to abnormal circumstances.

4.3.9 The handling of liquid hydrogen in transportable containers (not exceeding approx. 100 litres)

4.3.9.1 Use of correct containers

Only containers specifically designed to hold liquid hydrogen or approved for liquid-hydrogen service by the manufacturer should be used. Such containers are made from materials which can withstand the rapid changes and extreme differences in temperature encountered in working with the liquid. They should, however, be filled as slowly as possible, to minimize the thermal shocks which occur when any material is cooled.

Quantities of liquid hydrogen considerably greater than 5 litres should always be handled in enclosed vessels equipped with suitable relief valves or vents. Smaller volumes of liquid hydrogen may be handled in open-mouthed Dewar vessels. Such vessels should be stopped with as small an opening to the atmosphere as is consistent with the work to be done. All containers of liquid hydrogen should be vented or protected by a safety device which permits the escape of vapour but excludes entry of air. The vent should be checked at regular intervals to ensure that it does not become plugged with ice. Inadequate vent capacity can result in excessive gas pressure, which may damage or burst the container. It is recommended that, to prevent air ingress, containers are not completely emptied before being returned for refilling.

4.3.9.2 Preparations before filling

Before a warm container is first filled with liquid hydrogen, it should, if possible, be pre-cooled with liquid nitrogen. (Consult the container manufacturer for maximum weights of liquid nitrogen which may be used.) Pre-

Figure 12 Liquid hydrogen storage tank

cooling with liquid nitrogen will remove all the air from the container. It will also minimize the flash-off when the container is filled with liquid hydrogen. It is most important, however, that all the liquid nitrogen used for pre-cooling be removed from the container before adding hydrogen. If liquid nitrogen is not readily available, the container may be purged with gaseous nitrogen to make sure that all air is removed. After thorough purging with nitrogen, hydrogen gas should be used to displace the nitrogen, so that no nitrogen will freeze in the container when liquid hydrogen is introduced. In many cases this is achieved by boiling-off the first few cubic centimetres of liquid hydrogen introduced into the vessel.

Alternatively, the container may be purged by vacuum, using a suitable vacuum source, e.g. a hydraulically driven vacuum pump, provided that the container and its attachment can withstand the extreme pressure.

For filling containers previously used in liquid hydrogen service, purging is not required if there is positive assurance that the container holds only uncontaminated hydrogen. If it is suspected the container has become contaminated with air or other harmful impurities, the remaining liquid hydrogen should be removed. The container should then be allowed to warm sufficiently to vaporize any collected impurities, and then be purged with nitrogen gas. As with initial filling, the nitrogen purge should then be displaced with hydrogen gas; or vacuum purging may be used, if appropriate.

4.3.9.3 Use of correct transfer equipment

Use transfer equipment which has been designed for liquid hydrogen service. Unlike most other liquefied gases, liquid hydrogen should not be poured from one container to another or transferred in an atmosphere of air. If this is done, oxygen from the air will condense into the liquid hydrogen, adulterating it and presenting a possible explosion hazard. Pressurized withdrawal through an insulated tube is recommended. The liquid should be pressurized with very pure, dry, regulated hydrogen or helium only – not with air or nitrogen. Dewar flasks or other equipment made of glass should be used with caution because of the risk of breakage.

4.3.9.4 Handling of portable liquid hydrogen containers

Containers must always be handled and stored in an upright position and used in accordance with manufacturer's instruction.

Bibliography

Chelton D.B., 'Safety in the use of liquid hydrogen', ch. 10 in Scott, R.B., Denton, W.H., and Nicholls, C.M. (eds), *Technology and Uses of Liquid Hydrogen*, Pergamon Press, New York, 1964.

Connolly, W.W., 'A practical safety standard for commercial handling of liquefied hydrogen', *Advances in Cryogenics Eng., 12*, 192, *Proc. 1966. Cryo. Eng. Conf.*, 1967 Plenum Press NY 1967.

European Industrial Gases Association, Brussels. 'Gaseous hydrogen stations'. Industrial Gases Council Doc. 15/80

European Industrial Gases Association, Brussels. 'Liquid Hydrogen Storage Installations at users' Premises'. Industrial Gases Council Doc. 1990.

Himmelberger, F., 'Liquid Hydrogen Safety', Air Prod. Inc., Allentown. Pa., *Proc. 1959 Safety Conf.*, July 1959, p.49.

Hord, J., 'Is hydrogen safe?', NBS Technical Note 690, National Bureau of Standards, Boulder, Colorado, 1976.

Kerr, E.C., 'Liquid Hydrogen – a guide for the safe handling and storage of liquid hydrogen at LASL facilities', Los Alamos Sci. Lab., Liquid Hyd. Safety Com., N. Mex., Rep. No. 0-20, p.1, June 1962.

'NFPA 50B, 'Standard for Liquefied Hydrogen Systems at Consumer Sites' 1989 (National Fire Protection Association, Boston, Ma.).

'Sax, N.I., *Dangerous Properties of Industrial Materials,* Van Nostrand Rheinhold, New York, 1988.

Scharle, W.J., 'The safe handling of liquid hydrogen' *Chem. Engr.*, London, No. 185, Jan–Feb. 1965, p. CE-16.

Stoll, A.P., 'The storage and handling of hydrogen with safety' *Chem. Engr.*, London, Jan–Feb 1965, p.CE-11.

Van Meter, R.A., 'Some problem areas in Hydrogen Safety', Comp. Gas Assoc., Sppl. Ann. Rep., Nos. 51, 52. 1963-4, p.20.

Van Meter, R.A., Liebman, I. and Litchfield, E.L., *et al.*, *Hydrogen Safety*. US Bureau of Mines, Explosives Res. Center, Pittsburgh, Pa., Prog. Rep. No.3, October 1964, p.1.

Weintraub, A.A., 'Control of liquid hydrogen hazards at experimental facilities – a review'. Health and Safety Lab., NY, Operations Office, AEC Report No. HASL-160, May 1965, p.1.

Zabetakis, M.G. and Burgess, D.S., 'Research on the hazards associated with the production and handling of liquid hydrogen', Proj 8, (22-3151), June 1960. p.1.

Zabetakis, M.G. and Van Meter, R.A. *et al. Hydrogen Safety*. US Bureau of Mines, Explosives Res. Center, Pittsburg, PA. Prog. Report No.2, June 1964, p.1.

Zabetakis. M.G., Van Meter, R.A., Liebman, I. and Van Dolah. R.W., *Hydrogen Safety*. Bureau of Mines, Explosives Res. Center, Pittsburg, Pa., *Explos. Res. Center, Prog. Rep.* No 1, April 1964, p.1.

5 Helium and other rare gases – research systems

Chapter 5 deals with special precautions which must be observed in the handling of helium, neon, krypton and xenon. This section supplements the general safety requirements of Chapter 1.

5.1 Helium, neon, krypton and xenon

Liquid helium and the liquid phases of rare gases such as neon, krypton and xenon are usually encountered in much smaller quantities than other cryogens. However, the nature and uses of these products demand special care and procedures in their handling. This is particularly the case in small scale and research systems, where the application is such that only a very limited amount of prior experience may be available. For example, reference to Table 1 will show that helium has the lowest boiling point of any substance, $-269°C$ (4.2K) at 1 bar abs, while the only other substance with a lower boiling point than neon, $-246°C$ (27K) at 1 bar abs, is hydrogen.

The extremes of density associated with these substances should also be noted. Helium is the second lightest of the elements with a liquid density of $125 kg/m^3$ at its normal boiling point. By contrast, neon, krypton and xenon are the heaviest of the cryogens, with liquid densities of $1206 kg/m^3$, $2400 kg/m^3$ and $3040 kg/m^3$ respectively.

Taking the extreme case, liquid xenon is therefore more than twenty-four times as dense as liquid helium, and almost three times the density of liquid nitrogen.

5.1.1 Applications

Products such as neon, krypton and xenon, although produced and stored as cryogens, will ultimately be used in their gaseous form at ambient temperatures, the main application being in lasers, discharge tubes and other light emitters. However, all cryogens could be employed as refrigerants if required. In practice only nitrogen and helium are commonly used, other cryogens being used as refrigerants only in special circumstances, usually where intermediate temperatures are required. For example, the normal boiling point of neon lies conveniently between that of helium and nitrogen, and neon offers a safe alternative to hydrogen. Nitrogen has the advantages of being readily available, inexpensive and inert, and is used in a wide range of industrial, medical and food-processing applications. Helium is expensive but offers the lowest temperature of any of the cryogens. Large-scale use of helium, greater than a few litres, is limited to the cooling of superconducting magnets.

5.1.2 Potential hazards

The potential hazards in handling liquid helium, neon, krypton and xenon are similar:
(a) The liquids are extremely cold.
(b) The very low temperatures of helium and neon can condense and possibly solidify other gases. This may be a particular hazard in the case of uninsulated or poorly insulated transfer lines.
(c) The volume increase from liquid to vapour is considerable and may lead to high pressures in unvented enclosures.
(d) Low temperature embrittlement of materials may occur, particularly at helium and neon temperatures.
(e) These substances will not support life.
(f) The high densities of neon, krypton and xenon are such that Dewars designed for helium or nitrogen service are not usually suitable and may be unsafe.

Although the very low temperatures associated with helium and neon can present particular design and material problems, the procedures for safe handling of these rare gas cryogens are generally as outlined in Chapter 1. More specific information is given in the following sections.

5.1.3 Health and general safety

The precautions outlined in Chapter 1 of this manual concerning first aid, protective clothing, materials compatibility and fire and oxygen hazards apply to the handling of all cryogens, and should be followed. However, the nature of likely applications of these particular cryogens, perhaps including prototype or research systems, and the very low temperatures associated with helium and neon, demand that special care be taken.

5.1.4 Specific hazards

5.1.4.1 Condensation and solidification (helium and neon)

The temperature of liquid helium is sufficiently low to condense and freeze any other gas; neon is sufficiently cold to condense and freeze all gases with the exception of hydrogen and helium. Consequently, there is a danger of pipes or vents becoming plugged, and an air liquefaction hazard. Liquid helium and liquid neon must therefore be stored and handled under positive pressure or in closed systems to prevent the infiltration and solidification of air or other gases. As explained in Chapter 1, a plugged Dewar or storage vessel may develop sufficient internal pressure to cause catastrophic failure.

5.1.4.2 Oxygen enrichment (helium and neon)

Since air will condense on helium- or neon-cooled surfaces, there is a potential oxygen-enrichment hazard. At its dew point liquid air contains about 50 per cent oxygen by volume, and the oxygen content of any accumulation of liquid air will rise still further as the lower-boiling-point nitrogen component evaporates. Surfaces on which condensation might occur must be cleaned to oxygen service standards and must be kept free of oil, grease or other combustible material. Ideally, all cold surfaces should be vacuum-insulated, but if a solid insulant is used at any point, this should be oxygen-compatible and non-combustible.

5.1.4.3 Materials compatibility (helium and neon)

Many of the materials qualified for nitrogen service may not be suitable for use with liquid helium or liquid neon, because of their significantly lower temperatures. Care must therefore be taken to ensure that even equipment designed for cryogenic use is in fact suitable for the extremely low temperatures associated with these cryogens.

5.1.5 Handling equipment

5.1.5.1 Supply system design

Because of their limited quantities, cryogens such as helium, neon, krypton and xenon are conveniently transported and stored in small Dewars or liquid cylinders with capacities up to about 200 litres. Depending on the application, the product may be drawn directly from the main storage vessel or decanted into a smaller Dewar for subsequent use. It should be appreciated, however, that each transfer process will incur a loss of product; hence the number of transfers should be minimized by appropriate design wherever possible. A typical storage Dewar and supply system is shown in Figure 13. If the product is to be used as a refrigerant, a recovery system will be added if the quantities are significant.

5.1.5.2 Precautions when handling Dewars

In addition to observing the general safety considerations relating to Dewars and insulated vessels outlined in Chapter 1 it must be appreciated that the Dewars used to contain helium and the other rare gases may demand special care in handling. Dewar vessels are designed and constructed to minimize heat inleak, and, while not exactly delicate, do need to be treated with care. Most of the internal support is built into the neck, and the dimensions of all members linking the inner and outer vessels will be a minimum in order to reduce conduction losses. It is therefore important not to subject Dewars to mechanical shock or, in the case of certain designs, use them in other than the vertical position, especially when full. A 100 litre Dewar full of krypton would weigh in excess of 300kg. Dewars designed for helium service should not be used for any other purpose. Since the density of liquid helium is low, the vessel may not be adequately stressed to contain other cryogens.

Helium and other rare gases – research systems 97

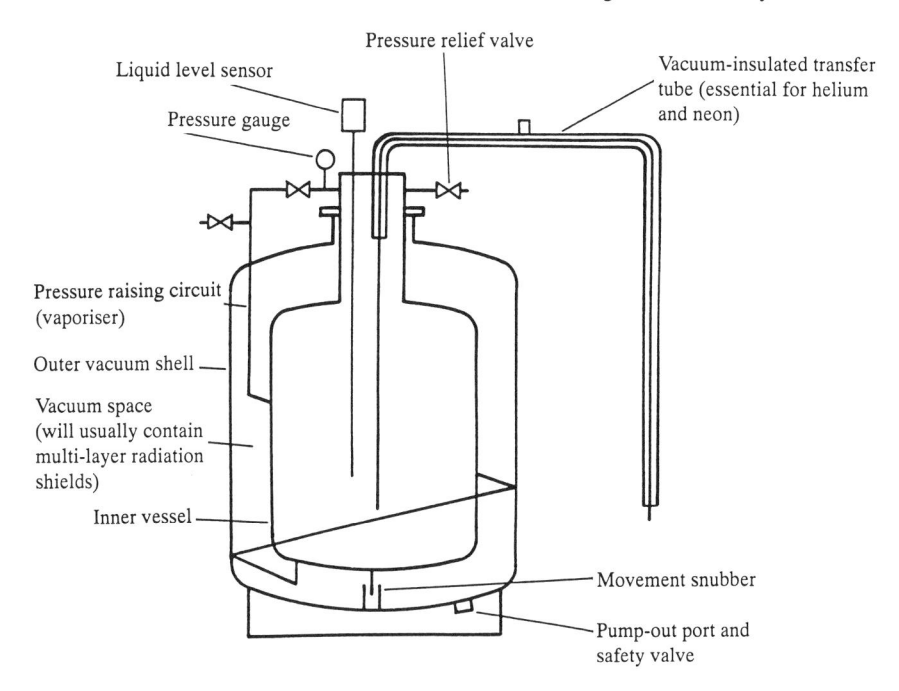

Figure 13 Typical liquid storage Dewar vessel and transfer tube

Dewars and many other storage vessels are designed to work over a very limited range of pressures. Care must be taken to ensure that overpressurization is avoided. Effective safety vents must be fitted and operational to minimize this risk. In all cases the manufacturer's operating instructions must be followed.

5.1.5.3 Transfer processes

It is good practice to transfer all liquid cryogens through insulated lines. Non-insulated transfer lines will result in loss of liquid and may give rise to problems associated with condensation of air and other gases. The use of vacuum-insulated lines is essential for helium transfer and desirable for other cryogens. If liquid loss is to be minimized, transfer lines should be pre-cooled with liquid nitrogen.

Before liquid transfer, all lines and receiving vessels should be purged with dry gas. Particular care must be taken to purge joints before assembly, and ensure that no foreign gases remain in closed-ended areas of a system.

Pressure for transfer may be provided by pressure build-up due to natural heat leak, a pressure raising circuit, by using a pressurizing bladder, or by supply of pure gas from a regulated cylinder.

In most cases low pressure transfer is to be preferred. The exit of a transfer line should be below the final liquid level. The cryogen should be directed towards the bottom of the receiving vessel to avoid excessive boiling and splashing. Where possible, the insertion of the transfer line should be delayed until cold vapour issues from the outlet, in order to avoid a build-up of back pressure in the line during cool-down.

If oscillations occur during transfer, the process should be stopped. This phenomenon of thermo-acoustic oscillation results from a particular combination of temperature gradient, heat transfer, system configuration and fluid properties. The oscillations are most often observed in helium transfer lines, where they can lead to a serious loss of liquid and, in extreme cases, even mechanical damage. The problem can usually be overcome by repositioning the transfer tube within the warm receiving vessel. Persistent oscillations may be eliminated by changing the volume of the receiving vessel. The phenomenon may be usefully employed in a liquid level sensing device.

5.2 Liquid helium

Liquid helium has many unique properties, and in particular has the lowest boiling point of all the cryogens. Helium gas is less dense than air and is present in the atmosphere in the ratio of about five parts per million. Practically all commercial helium, however, is obtained from natural gas, where it is typically found in concentrations of between 0.5 and 2.5 per cent by volume, although particular gas reserves may contain as much as 7 per cent helium.

Helium exists as several isotopes, the most common being He^4. The next most abundant isotope is He^3, although this is rarely found outside the research laboratory (ordinary helium gas contains only about 1.3×10^{-4} per cent He^3). Whenever helium is referred to without any isotope designation it can be assumed to be He^4. Mixtures of He^3 and He^4 are used in dilution refrigerators to reach temperatures below $-272°C$ (1K).

In liquid form helium exhibits two very different characteristics, depending on temperature (Figure 14). Between the critical temperature of $-268°C$ (5.2K) and the lambda line it is known as liquid helium I, and apart from its very low temperature exhibits characteristics typical of many liquids. However, at the lambda line the liquid undergoes a transition (the lambda transition) and at lower temperatures becomes helium II. This colder liquid exhibits the phenomenon of superfluidity, having virtually zero viscosity, and in addition has an extremely high effective thermal conductivity (over 1000 times as great as that of copper) near the lambda line.

Among the various applications for liquid helium are the cooling of superconducting magnets, space simulation and the production of ultra high vacuum. Although high concentrations of gaseous helium are used in certain breathing gas mixtures by divers, in its pure form it will not support life and must therefore be treated as an asphyxiant. In general, the properties of liquid helium are sufficiently unusual to demand special equipment and handling techniques.

5.2.1 Vaporization of liquid helium

Owing to its relatively low latent heat of vaporization (less than 1/50th that of liquid nitrogen) liquid helium will evaporate rapidly when heated or when

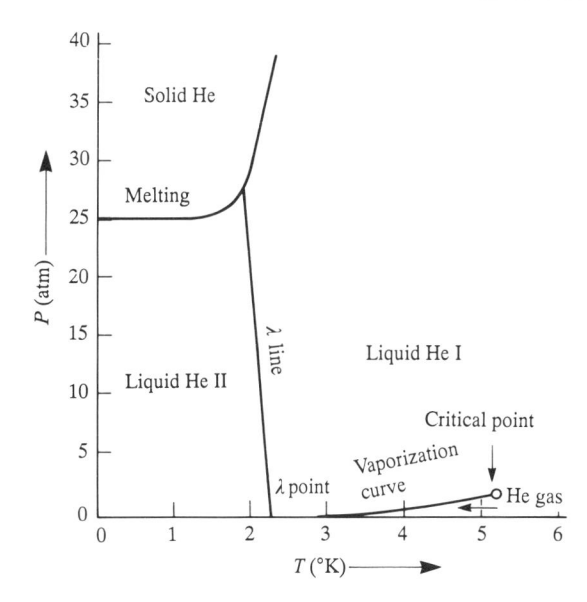

Helium and other rare gases – research systems 99

Figure 14 Phase diagram for helium

liquid is first transferred into warm or partially cooled equipment. Similarly, only a very limited deterioration in the insulating vacuum of a storage vessel will result in a large increase in the rate of boiling of the liquid. Pressure relief devices for liquid helium systems must therefore be of adequate capacity to release the large quantities of vapour that may result under such circumstances.

5.2.2 Special precautions with helium Dewars

The very low density of liquid helium is such that helium Dewars may be of particularly light internal construction, thus minimizing conduction paths and thermal mass. It should be emphasized that this mechanical specification does mean that helium Dewars should never be used for other cryogens, since this could result in a failure of the internal suspension system. For similar reasons it is particularly important that helium Dewars are not tilted or subjected to mechanical shock, especially when full. During transfer operations care must be taken to avoid overpressurization.

5.3 Grouping of neon, krypton and xenon

It is appropriate that neon, krypton and xenon should be grouped together, since these products are used in similar applications, at least in their gaseous forms. The gases are produced commercially as a by-product of air liquefaction, although the yield is small because they are only present in the atmosphere in very low concentrations. The cooling capacity of liquid neon is forty times that of liquid helium per unit volume, and more than three times that of liquid hydrogen. Further, neon has a higher latent heat capacity than helium, and, unlike hydrogen, is non-flammable. Despite these desirable properties, its use as a refrigerant is still limited. Other than in very special circumstances it is unlikely that krypton or xenon would be used as refrigerants.

The normal boiling point of neon ($-246°C$, 27K) is considerably lower than that of krypton ($-153°C$, 120K) or xenon ($-108°C$, 165K), and as such imposes more stringent requirements on system design and operation. In general, the procedures for handling liquid neon are similar to those for helium, while for krypton and xenon procedures used for nitrogen are appropriate. Neon, krypton and xenon are not in themselves harmful, but all will act as asphyxiants by displacing oxygen.

5.3.1 Thermophysical properties

The following data supplement that in Table 1.

Neon   Abundance in dry air 15.5ppm.
        Three stable isotopes.
        Inert (under most conditions).
        Non-toxic.
        Asphyxiant.

Krypton Abundance in dry air 0.14ppm.
        Six stable isotopes.
        Inert (under most conditions).
        Used to define SI standard of length.
        Non-toxic.
        Asphyxiant.

Xenon   Abundance in dry air 0.9ppm.
        Nine stable isotopes.
        Often considered inert but rare compounds are possible
            which may be toxic and even explosive. However, these
            are unlikely to be encountered in cryogenic plant.
        Non-toxic in pure form.
        Asphyxiant.

Bibliography

Barron, R.F., *Cryogenic Systems*, Oxford Universtiy Press, 1985.
Van Sciver S.W., *Helium Cryogenics*, Plenum, 1986.
Wigley D.A., *Mechanical Properties of Materials at Low Temperature*, Plenum, 1986.

Index

Acetylene, 32, 34
  testing apparatus, 35
Air:
  condensation, 79
  liquid:
    burns, 89
    disposal, 38
Air separation plant, 30–49
  liquid nitrogen decanting into, 35–6
  liquid oxygen decanting into, 35–6
  maintenance, 45–50
    acid cleaning, 48
    caustic cleaning, 48
    cleanliness, 45–6, 47
    cleaning methods, criteria of selection, 47
    cleaning processes, 48–9
    drying, 49
    facilities, 46
    equipment, 45
    inspection techniques, 47
    materials, 50
    mechanical cleaning, 48
    packaging, 49
    personal cleanliness, 46
    precautions, 49
    protective clothing, 46
    rinsing, 49
    special area, 46
    solvent washing, 48–9
    spares, 50
    steam/hot water cleaning, 48
    vapour degreasing, 48–9
    work benches, 46
  modifications, 39
  location, 32
  operation, 30
    smoking/open flame prohibited, 30
  purging, 33
  vents, 38
Alloy steels, 9
Aluminium, 42
  coefficients of expansion, 12 (table)
Ambient air vaporizers, 38

Argon:
  hazards, 29–30
  leaks, 37–8
  thermophysical properties, 2 (table)
Aromatic hydrocarbons, 81
Asphyxiation hazard warning sign, 6 (fig.)
Autostratification, 70

Barrier protection cream, 46
Benzene, 63
Brass:
  alloys, 42
  coefficients of expansion, 12 (table)
Brelite (perlite), 49
Brittle failure, 9, 10 (fig.)
Burns, 1–3
  first aid, 3
  precautions, 1–3

Carbon, 9
Carbon dioxide, 32
  removal by purging, 55
Carbon monoxide, 82
Carbon steel, oxygen-clean, 42
Catalytic filters, 33
Centrifugal oxygen compressor, 39–40
Charpy V-notch impact tests, 9 (fig.)
Chlorinated solvents, 49
Coefficients of expansion, 12 (table)
Cold box purge, 36–7
Column sump liquid, 35
Condensation of gases, 79
Composites, 9
Contaminants in process stream, 32–6
  airborne, 32–3
  air separation plant purging, 33
  catalytic filters, 33
  column sump liquid, 35
  construction introduced, 36
Control of plant regenerator/heat exchange temperatures, 33

102   Index

Control of plant regenerator/heat exchange temperatures – *continued*
  decanting of liquid oxygen/liquid nitrogen, 35–6
  drainage of plant liquid after shutdown, 35
  equipment introduced, 36
  hydrocarbon absorbers, 33
  liquid level control, 33
  maintenance introduced, 36
  molecular sieve adsorbers, 33–4
  monitoring, 34–5
  plant de-rime (defrost) schedules, 35
Contraction, 11
Control of Industrial Accident Hazards Regulation 1984, 25–6
Copper alloys, 42
Cryogens:
  definition, 1
  distribution, 15–18
  leakage, 11
  liquid, disposal, 38
  rapid changes in operating parameters, 24
  thermophysical properties, 2 (table)
  toxicity, 5–6

Dangerous Substances (Notification and Marking of Sites) Regulation 1990, 26
Defrost (plant de-rime) schedules, 35
Detection systems, 74–5
  fire, 75
  flammable gas, 74
Dewars, 16–17, 96–7
  liquid helium, 99
Diesel-engined vehicles, 65
Drains, 65

Elastomers, 10
Ethane, 52–75
  analyser house, 57
  asphyxiation, 56
  compressor house, 57
  control room, 57
  confinement effect on fire, 53
  electrostatic charges, 55
  flame characteristics, 54
  flammable range, 53
  health hazards, 56
  laminar burning velocity, 52
  limiting oxygen index, 53
  minimum ignition energy, 53
  overpressure, 54
  oxidants, 54–5
  safety in maintenance, 60–66
    entry to vessels/confined spaces, 63–5
    hot work, 65–6
    isolation, 61–2
    permits to work (clearance certificates), 20–23 (fig.), 61, 62, 63–4
    responsibility, 60
    routine safety maintenance frequency, 60–61
    sweeping out, 62–3
    testing, 63
  safety in operation, 56–60
  secondary hazards, 56
  small storage installation, 58–9
  spontaneous ignition temperature (SIT), 52
  thermophysical properties, 2 (table), 53 (table)
  ventilation of plant building, 57
Ethylene, 52–75
  analyser house, 57
  asphyxiation, 56
  compression house, 57, 67 (fig.)
  confinement effect on fire, 53
  control room, 57
  electrostatic charges, 5
  flame characteristics, 54
  flammable range, 53
  health hazards, 56
  laminar burning velocity, 52–3
  limiting oxygen index, 53
  minimum ignition energy, 53
  overpressure, 54
  oxidants, 54–5
  quenching distance, 54
  safety in maintenance, 60–66
    entry into vessels/confined spaces, 63–5
    hot work, 65–6
    isolation, 61–2
    permits to work (clearance certificates), 20–23 (fig.), 61, 62, 63–4
    responsibility, 60
    routine safety maintenance frequency, 60–61
    sweeping out, 62–3
    testing, 63
  safety in operation, 56–60
  secondary hazards, 56
  small storage installation, 58–9
  spontaneous ignition temperature (SIT), 52
  storage tank, 67–8
    autostratification, 70
    rollover, 69–70
  thermal decomposition, 54
  thermophysical properties, 2 (table), 53 (table)
  ventilation of plant building, 57
Ethylene/diatomic gas mixture, 54
Explosion hazard *see* Fire/explosion

Factories Act 1961, 4
Fatigue failure, 9, 11
Fire/explosion, 8–9
  hydrogen, 77–80
  detection, 75
  duration, 71

Index 103

extent, 71
flammable leaks/delayed ignition, 71
high-pressure storage, 72
initiation, 70–71
natural gas, ethane, ethylene, 52–4
confinement effect, 53
decomposition, 54
flame characteristics, 54
flammable range, 53
laminar burning velocity, 53–4
limiting oxygen index, 53
major leaks, 71–2
minimum ignition energy, 53
minor leaks, 71
quenching distance, 53–4
spontaneous ignition temperature (SIT), 52
thermal expansion, 11–13, 39, 54
oxygen enrichment, 9
Firefighting:
hydrogen, 80
natural gas, ethane, ethylene, 55–6, 72–4
equipment, 73–4
foams, 72–3
suits, 73
Fire triangle, 8 (fig.)
Frostbite, 1–3
Flammable gas detectors, 13–14

Gas detection, 13–15

Helium, 94–9
application, 94–5
boiling point, 94
condensation, 95–6
Dewars, 96–7
health/safety, 95
isotopes, 98
liquid, 98–9
density, 94
Dewars, 99
vaporization, 98–9
materials compatibility, 96
oxygen enrichment, 96
solidification, 95–6
supply system design, 96
thermo-acoustic oscillation, 98
thermophysical properties, 2 (table)
transfer process, 97–8
Hot work permit, 65–6
Hydrocarbons, 32
absorbers, 33
desorbed, 34
Hydrogen, 77–92
boiling point, 94
confinement effect on fire, 77–8
electrical equipment, 80
electrostatic effects, 79–80
fire/explosion hazard, 77–8

firefighting, 80
flame characteristics, 78
flammable range, 78
health hazards, 81
laminar burning velocity, 77–8
leakage, 11
limiting oxygen index, 78, 79
liquid *see* Liquid hydrogen
mechanical hazard, 81
minimum ignition energy, 78
quenching distance, 78
spontaneous ignition temperature (SIT), 77
thermophysical properties, 2 (table)
Hydrogen plants, 82–8
cold box, 82–3
plant external to, 83
Hydrogen separation/liquefaction plant, 83–8
safety in maintenance, 83–6
cleanliness, 83
equipment isolation, 84
ignition sources, 84–5, 85–6
personal cleanliness, 84
protective clothing, 84
purging of vessels/confined spaces, 84
safety control procedures, 84
special area, 83
tools, 84
ventilation, 84
work permit area, 85
safety in operation, 86–8
cold box atmosphere, 87–8
disposal of liquefied gases, 88
emergency procedure, 88
fire danger, 86
heat danger, 86
leaks, 86–7
naked lights prohibited, 86
purging before start-up, 88
safety devices, 87
smoking prohibited, 86
unsafe materials, 86
venting of trapped liquefied gases, 88
Hydrogen sulphide, 81–2

Ice formation, 11
Impact energy (toughness), 9 (fig.)
Insulated vessels, 16–17
Invar, coefficients of expansion, 12 (table)

Krypton, 94–100
application, 94–5
Dewars, 96–7
grouping, 99–100
hazards, 95
health/safety, 95
liquid density, 94
supply system design, 96
thermo-acoustical properties, 2 (table), 100

Krypton – *continued*
    transfer processes, 97–8
Legislation, 25–6
    Control of Industrial Major Accident
        Hazards Regulations 1984, 25–6
    Dangerous Substances (Notification and
        Marking of Sites) Regulations 1990,
        26
    EC Council Directive, 25
    Factories Act 1961, 4
    Notification of Installations Handling
        Hazardous Sustances Regulations
        1982, 25
    planning controls, 26
Liquid hydrogen, storage, transport,
    handling equipment, 89–93
    construction materials, 90
    identification, 89
    loading/transfer areas, 91
        access, 90
        no smoking area, 90
        protective clothing, 90
        purging, 90
        storage/loading area, 89
        tow-away accident, 90
        transfer lines, 89
    transportable containers, 91–2
    ventilation, 90
Limiting oxygen index:
    ethane, 53, 54
    ethylene, 53, 54
    hydrogen, 78
    methane, 53, 54
Liquid cylinders, 17–18
Liquid oxygen pumps, 41
Liquid storage containers, 42–5
    ancillary equipment failure, 44
    overfilling, 43
    overpressure, 43
    purge gas flow failure, 44
    underpressure, 43–4
Low-pressure plants, 36
Lubricants, 36

Mechanical properties of materials, 9–11
Methane:
    limiting oxygen index, 53
    thermophysical properties, 2 (table), 53
        (table)
    see also Natural gas
Molecular sieve adsorbers, 33–4

Natural gas, 52–75
    analyser house, 57–8
    asphyxiation, 56
    compressor house, 57
    confinement effect on fire, 53
    control room, 57
    electrostatic charges, 55
    flame characteristics, 54

flammable range, 53
health hazards, 56
laminar burning velocity, 52–3
minimum ignition energy, 53
overpressure, 54
oxidants, 54–5
quenching distance, 54
safety in maintenance, 60–66
    entry to vessels/confined spaces, 63–5
    hot work, 65–6
    isolation, 61–2
    permits to work (clearance certificates),
        20–24 (fig.), 61, 62, 63–4
    responsibility, 60
    routine safety maintenance frequency,
        60–61
    sweeping out, 62–3
    testing, 63
safety in operation, 56–60
secondary hazards, 56
small storage installation, 58–9
spontaneous ignition temperature, 52
storage, 66–70
ventilation of plant building, 57
see also Methane
Neon, 94–100
    applications, 94–5
    boiling point, 100
    condensation, 95
    Dewars, 96–7
    grouping, 99–100
    hazards, 95
    health/safety, 95
    liquid, 99–100
        density, 95
    oxygen enrichment, 96
    solidification, 95
    supply system design, 96
    thermo-acoustic oscillation, 98
    thermophysical properties, 2 (table), 100
    transfer process, 97–8
Nickel alloys, 42
Nickel carbonyl, 82
Nitrogen:
    hazards, 29–30
    liquid:
        decanting into air separation unit, 35–6
        disposal, 38
        purge, 54
    thermophysical properties, 2 (table)
Nitrogen compressors, 41–2
Nitrogen oxides, 32, 79
Notification of Installations Handling
    Hazardous Substances Regulations
    1982, 25

Openpath infrared sensors, 75
Oxygen, 27–9
    condensation, 79
    deficiency (anoxia), 3–5

hazard warning sign, 6 (fig.)
detection, 13
fire enrichment, 9
fires in oxygen service, 42
hazards, 27–9
  to personnel, 28–9
leakage, 11, 37–8
limiting index, 53
liquid:
  decanting into air separation unit, 35–6
  disposal, 38
liquid pumps, 41
misuse, 29
pipelines, 42
reaction with oil/grease, 45
selection of materials for service, 27–8
smoking/open flame forbidden, 30
thermophysical properties, 2 (table)
vapour cloud, 38
Oxygen compressor:
  centrifugal, 39–40
  reciprocating, 40–41
Oxygen turbocompressors, 46

Paramagnetic analysers, 13
Perlite (brelite), 49
Petrol-engined vehicles, 65
Pipelines, 15, 42
Plant de-rime (defrost) schedules, 35
Polymers, 9
Pressure relief valve, 39, 54
Pressure vessels registration cards, 60
Protective clothing/equipment, 24
PTFE, coefficients of expansion, 12 (table)
Pyrex glass, coefficients of expansion, 12 (table)

Rail tanker, 15, 17 (fig.)
Reboiler-condenser operation, 34
Reciprocating machinery, 36
Reciprocating oxygen compressor, 40–41
Road tanker, 15, 16 (fig.), 44–5, 58
  earth cable, 55

tow-away incident, 44, 90
Rollover, 69–70
Rust particles, 42

Safe system of work, 4–5
Safety control procedures, 18–24
  actions to make plant/equipment safe, 19
  permits, 18–19, 20–23 (fig.)
  protective clothing/equipment, 24
    special precautions, 24
Safety devices, 7
Stainless steel, 42
  coefficients of expansion, 12 (table)
Stress cracking, 9
Sulphur dioxide, 32

Thermal expansion, 11–13, 39, 54
Thermo-siphon reboilers, 34
Toughness (impact energy), 9 (fig.)
Toxic gas detectors, 14–15
1,1,1, trichloroethane, 48
Trichloroethylene, 48

Vapour fog clouds, 38–9
Vehicle entry into plant areas, 59–60
  high-risk area, 60

Xenon, 94–100
  application, 94–5
  Dewars, 96–7
  grouping, 99–100
  hazards, 95
  health/safety, 95
  liquid density, 94
  supply system design, 96
  thermo-acoustic oscillation, 98
  thermophysical properties, 2 (table), 100
  transfer processes, 97–8

Zirconia analysers, 13